MATERIAL CONCERNS

Pollution, profit and quality of life

Tim Jackson

London and New York

First published 1996
by Routledge
11 New Fetter Lane, London EC4P 4EE

Simultaneously published in the USA and Canada
by Routledge
29 West 35th Street, New York, NY 10001

Routledge is an International Thomson Publishing company

Typeset in Bembo by
Florencetype Ltd, Stoodleigh, Devon

Printed and bound in Great Britain by
Redwood Books, Trowbridge, Wilts

British Library Cataloguing in Publication Data
A catalogue record for this book is available from the British Library

Library of Congress Cataloguing in Publication Data
A catalogue record for this book has been requested

ISBN 0–415–13248–7 (hbk)
ISBN 0–415–13249–5 (pbk)

CONTENTS

PLATES

FIGURES

BOXES

PREFACE

As the title suggests, this book is mostly about the material basis of the world in which we live. Over the course of the last two centuries, that basis has undergone some fairly radical changes which, taken together, constitute a revolution in the relationship between human society and its natural environment. This is especially true of the so-called 'developed' nations which have built considerable wealth from advanced industrial economies. But industrialisation is also the aspiration of almost every other country in the world. So the evolution witnessed during this period of extraordinary change is really global in extent.

Industrialisation has not come without a price, however. It relies on continued access to limited material resources. It imposes increasingly demanding burdens on the environment. These are – in part – the material concerns to which the title of this book alludes. The prospects of global warming, ozone depletion, pollution of water supplies, soil degradation, deforestation, desertification (and so on) have haunted the progress of civilisation and now threaten to undermine economic development. Many people acknowledge the urgency of these problems and the need for action. But there is a fundamental division over the appropriate response.

Some point to the enormous benefits which the industrial economy has provided: increased life expectancy, reduced manual drudgery, better education, and technological advances in health care, transport and communications. They see economic growth as crucial for the translation of these benefits to the global population and the development of new technological solutions to environmental problems. They argue that wealth is critical to the improvement of environmental performance.

Others insist that the pursuit of economic wealth and the indomitable rise in material expectations are the root cause of environmental degradation. They point to the increased material throughput of the industrial economies and the disproportionate burden placed on limited resources by the wealthy nations. They believe that radical measures are needed to curb environmental emissions and restrain economic expansion.

The idea that environmental protection and economic development are in natural opposition to one another has emerged as a common assumption, almost a defining characteristic of the dispute. One side has used this assumption to argue for better environmental protection and reduced economic activity. The other has used the same assumption to argue for increased economic activity with which to pay for environmental protection.

Two elements within a complex dialogue are beginning to disturb this well-established division of views.

The first is the suggestion that the industrial economy is showing signs of internal stress. Saturation of Northern markets, the growth of global competition, and systemic rises in unemployment levels have troubled the developed nations during the latter part of the twentieth century. It no longer seems as clear as it used to do that an economic system predicated on continued growth is viable.

The second element is the arrival of what might be called 'preventive environmental management'. Proponents of this emerging approach insist that it is possible to protect the environment without jeopardising economic competitiveness. They cite evidence of firms which save money and reduce pollution simultaneously. They talk of redesigning industrial policy to benefit both the environment and the economy. The idea that you can profit from improved environmental performance has gradually eroded the simplicity of the early debate and offered the promise of new kinds of solution to the underlying conflict.

How far can this new approach take us? To what extent is it feasible to implement the necessary changes within the existing economic system? How will they affect the behaviour of companies, governments and individuals? Must we change the system itself to accommodate these new ideas? Does preventive environmental management impose limits which will eventually constrain our development? Most fundamentally, will these new strategies deliver an acceptable level of environmental protection, without jeopardising human welfare?

These are among the questions which I have set out to examine in this book. The task is a frighteningly complex one, partly because it needs to draw on a very wide knowledge base. So the first chapter in the book starts out in the realms of pure science: ecology, physics, thermodynamics. The middle ground over which the book travels is largely technological. In fact, I have attempted to relate the discussion to practical examples from start to finish.

On the other hand, this is not a traditional technical textbook. Economics must play an absolutely vital role in the reorientation of industrial society, because it has played an absolutely vital role in the development of that society. But history, philosophy and psychology also creep into the discussion with a kind of uncanny persistence, as the book develops. We are not simply technological creatures living in an advanced industrial economy. We are complex human beings enmeshed in an intricate historical and social framework. Realistic solutions to systemic problems will not be found without paying attention to the breadth and depth of that underlying framework.

A part of my aim in this book is to guide the reader through at least a part of this complex network of interrelated intellectual disciplines. Each of them is important to a full understanding of the problems facing us. Each of them has a role to play in our search for solutions. But this is not just a guidebook. In fact, I am seeking to convey a very particular thesis about the reinvention of the industrial economy.

The starting point for that thesis is a recognition that the environmental concerns of the late twentieth century are material to the future of the industrial economy, and possibly to the survival of the human species. Much of the book is dedicated to the search for practical ways of reducing the material impacts of human activities. Throughout the book, however, I am also attempting a specific critique of the economic model which drives this material system. Towards the end of the book I will present a vision of human development which provides something of an alternative to what many would regard as the prevailing wisdom. Perhaps ironically, this vision has at its heart the idea that we have placed an undue emphasis on the material dimensions of human society. Material concerns are not, at the end of the day, the limit of human experience.

Inevitably, not everyone will agree with the thesis I am presenting. Inevitably also, there are aspects of the discussion which I have not been able to accord the weight which perhaps they deserve. Nevertheless, it is my belief that the breadth and scope of the reflections in this book are vital to a successful solution of the environmental problems which face the world today.

Tim Jackson
University of Surrey, June 1995

ACKNOWLEDGEMENTS

This book has evolved from a research project carried out by the Stockholm Environment Institute into industrial environmental management. The aim of that project was to draw together experts in the field from all over the world, and to pool their technical and visionary expertise in such a way as to draw out a coherent and detailed picture of the new 'preventive' environmental strategy. A previous publication (Jackson, 1993) has collected together the individual technical contributions which furnished the basis for that task. This book attempts to provide an overview of the emerging vision, in a form which is accessible to a wider audience. Because it has been written by a single author, it is inevitable that this vision will remain to some extent personal. At the same time, I have attempted in what follows to synthesise some kind of consensus from the individual contributions to the SEI project and could not proceed without expressing my profound gratitude for the intellectual input of those who have contributed to that consensus.

My thanks are due to all those who participated in two international workshops on preventive environmental management, and contributed to the technical papers which formed the basis for this work. I am indebted to Dr Brian Wynne, Research Director at the Centre for the Study of Environmental Change in Lancaster University (CSEC) for his collaboration in the early stages of the project, and to all those at SEI who made that project possible: Professor Gordon Goodman, who provided the initial stimulus for action, Professor Michael Chadwick for his unwavering personal and intellectual support, Dr Lars Kristoferson for his expert advice and guidance, Arno Rosemarin and Heli Pohjolainen for their expertise in the publication department, Solveig Nielsson, Cecilia Ruben and Gertrud Wollin for invaluable help and support during the two international seminars.

I am grateful for the financial support of the Stockholm Environment Institute, the Engineering and Physical Sciences Research Council and the Royal Academy of Engineering during the preparation of this book. Thanks are also due to Sarah Lloyd and Matthew Smith at Routledge for their support and enthusiasm throughout, and to Penny Jackson for preparing the figures and diagrams.

I owe a particular debt of gratitude to those who have generously committed time and energy to critical discussions and reviews of the material in the book: Kate Burningham, Michael Chadwick, Julian Clift, Roland Clift, David Fleming, David Gee, Michael Jacobs, John Jackson, Mark Jackson, Richard Jackson, Joanne Jordan, Alex MacGillivray, Nic Marks, Colin Sweet and Jim Sweet.

1

LIVING IN A MATERIAL WORLD

Rough guide to a lonely planet

INTRODUCTION

We are living in a material world. To say this is not just to say that the affluent consumer societies of the Western world are excessively materialistic. It is not just to claim that our priorities and our values have become increasingly embedded in the ownership of material possessions. These claims may be true, and at a later stage of this book, I shall examine that possibility further. But there is something much more basic involved in saying that we live in a material world.

Many of our most vital needs are essentially material ones: food, water, shelter, clothing and fuel. We survive as human beings by cultivating crops to convert to foodstuffs, manufacturing textiles to turn into clothing, excavating clay, sand and rock to build homes for shelter, mining coal and oil and gas to provide us with warmth, light and mobility, and extracting metals from ores to make the machinery and appliances we need to do all this.

In fact, there is a sense in which life itself is a fundamentally material concern. All biological organisms require energy to maintain life. Some organisms (green plants) are able to obtain this life energy directly from the sun. Many organisms (including human beings) have to obtain life energy by feeding on other material organisms. The process of digestion converts food into faeces, and releases energy. This energy allows us to maintain our complex biological structure, to forage for food, to reproduce the species, and to defend ourselves against predators. Without these material inputs and outputs we simply could not survive. So to say that we are living in a material world is to say something fundamental about the interaction between human society and its environment.

These days, of course, the scale and complexity of our material interactions are vastly increased over those of earlier societies, and over those of other biological organisms. The material requirements of 'advanced' industrial societies extend far beyond the survival needs of food, warmth and shelter. There are now growing demands for a wide range of material goods from aerosols to aeroplanes, cosmetics to computers, and vinyls to videos.

In spite of this complexity, there are two aspects of the industrial economy which relate it directly to other more 'primitive' societies, and indeed to the social organisation of other biological species. The first aspect is the common aim of survival. The second is the common set of physical laws which govern behaviour in all material systems. This common physical basis is so critical to the interaction between the human species and its environment that we must gain some understanding of it, right at the outset, before we can proceed with the investigations. The aim of this first chapter is to provide that understanding.

A THUMBNAIL SKETCH OF THE INDUSTRIAL ECONOMY

A simplified picture (Figure 1) will help to place some elementary structure on the complexity of the industrial economy. It is clear from the diagram that there is a more or less linear flow of materials through the system. Material resources extracted from the environment at one end of this flow are processed in various ways to provide goods and services within the economy, before flowing out of the economic system back into the environment as emissions and wastes.

There is an important distinction between two different types of resource inputs. The first type of resource is called **renewable** resources. These resources are provided on a continuous basis by the flow of certain kinds of materials and energy through the environment in well-established cycles. Renewable resources include many timber and forest products and agricultural products of various kinds.

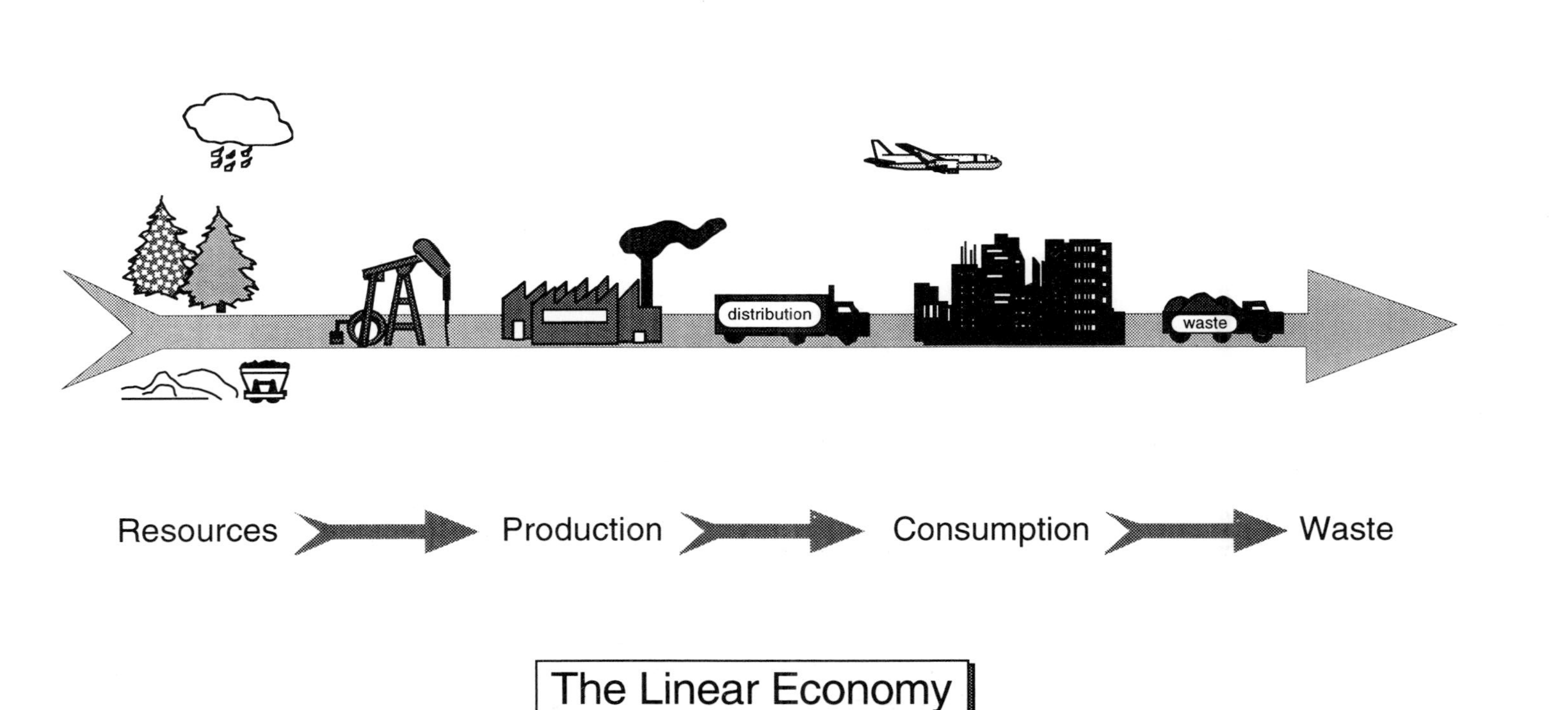

Figure 1 Material dimensions of the industrial economy

The second type of resource is **non-renewable**. We gain access to these resources only by depleting finite stocks of materials which have accumulated in various places in the environment sometimes over many thousands of years. The industrial economy now relies heavily on a number of non-renewable resources, including coal and petroleum, metallic minerals (iron, copper, aluminium, zinc, cadmium, mercury and lead, for instance), and various important mineral rocks such as phosphate, limestone and slate.[1]

Resource inputs (of both types) are then subject to various stages of processing and distribution in order to provide the goods and services demanded in the economy. The first stage of processing – often called primary processing – involves separating pure materials from the mixed form in which they are usually extracted from the environment. Physical, chemical and thermal separation and recombination processes convert the raw materials into finished materials such as fuels, refined metals and alloys, and industrial chemicals. Examples of such processes are the roasting and smelting of metal ores to separate metal from rock, the 'cracking' of crude petroleum to obtain specific oil derivatives, and the threshing of grain to separate wheat from chaff.

Materials purified in the primary sector are destined for the manufacturing industries, where materials are transformed into finished products. These industries are sometimes called the secondary sector of the economy. They include the textiles industries, the food-processing industries, the pharmaceuticals and cosmetics industries, the automotive industry, and the electronics industry.

Finished products from the secondary sector are then transported and distributed to the consumer. Most of the products are destined for another part of the industrial economy, sometimes called the tertiary sector. Part of this sector consists of retail trades which distribute the finished products to household consumers. The rest of the sector provides different kinds of services such as health, education, transport, household utilities, banking and communications.[2] Together with households, the tertiary sector is the main 'consumer' of the material products of the industrial economy. In addition, some of the products of the primary and secondary sectors are consumed within those sectors. For instance, the mining industry uses drilling and extraction machinery, the electricity supply industry needs combustion and generation technology and so on.

ENVIRONMENTAL IMPLICATIONS

When it comes to environmental impacts, the primary sector activities tend to be intrinsically 'dirtier' than secondary sector activities. The reason for this is quite straightforward. The role of primary sector processes is to separate and purify certain desired materials from a mixture of materials extracted from the environment. This means two things. First, the separation of pure materials from a mixture requires the input of energy. Second, the process of purification implies that some materials from the mixture are unwanted and must be discarded as wastes.

Consider the processing of pure copper metal from ores. To start with, excavating mineral ores requires the stripping of topsoils and rocks. This overburden can sometimes be several times greater than the weight of ore excavated. Next, the concentration of copper in copper ores is typically as low as 0.5 per cent to 2 per cent by weight. This means that for each kilogram of copper metal, between 50 and 200 kilograms of residue may be created. Finally, extracting copper from copper ore requires an energy input of between 50 and 100 megajoules[3] per kilogram of finished copper.[4] This separation energy is almost always provided by burning fossil fuels. And the combustion of fossil fuels is responsible for some of the most pressing environmental problems we face: global warming, acid rain, local air pollution, and the need to dispose of contaminated ashes and residues.

By contrast, the secondary sector activities tend to be less energy-intensive and often less liable to polluting residues, precisely because they are dealing with purer inputs. This does not mean, however, that they are pollution free. We shall see later in this chapter that any industrial process must inevitably generate wastes. Secondary sector industries are no exception.

The distribution and transportation of goods from manufacturers to retailers and then to consumers also gives rise to environmental impacts. In fact, the development of the industrial economy over the last fifty years has witnessed two trends in the development of transport requirements with specific environmental implications. In the first place, there has been a tendency towards centralisation – both of production facilities and of distribution outlets. Second, there has been an increasing globalisation of world trade. These two facets of the modern industrial economy impose special demands on the transport infrastructure. Goods must be transported often thousands of miles from the point of production to the point of consumption. Consumers often travel increasing distances from home to the point of sale. The increased demand for transportation has led to increased vehicle emissions, the loss of land to highways, railways, ports and airports, and the nuisance and public health effects which these developments bring with them.

The tertiary sector does not itself manufacture material products. Consequently, it does not generate production wastes. But we should not be seduced into believing that its activities are inherently clean. Most importantly, this sector consumes material products. And these products must be provided by the primary and secondary manufacturing industries. So in a sense, it is the demands of the tertiary sector which are partly responsible for the pollution from the primary and secondary sectors.

A similar thing can be said about households. In fact, householders create both the demand for material goods and the demand for services provided and distributed by the tertiary sector. This demand for goods and services is really the engine which is driving the industrial economy. It is, if you like, the root cause of the environmental impacts from the primary, secondary and tertiary industrial sectors and much of the transportation network. And this same demand generates increasing quantities of household and consumer waste: products and materials which are thrown away once they have reached the end of their useful life.

Some kinds of products are inherently **dissipative** in nature. Using them means consuming them in such a way that materials become widely dispersed into the environment. Some obvious examples of this kind of **dissipative consumption** are the combustion of fossil fuels, the use of domestic pharmaceuticals such as soaps, washing powders, detergents and bleaches, and the spreading of chemical fertilisers, herbicides, insecticides and fungicides. Some slightly less obvious examples include the use of lead as an additive in motor fuels, the use of zinc additives in rubber, which are released when tyres wear away, and the use of cadmium in metal plating. Other products are not *inherently* dissipated in this way. But the pattern of consumption of most products leads to their wide geographical dispersal throughout the economy. Disposal of these products at the end of their lives also gives rise to the dissipation of materials into the environment.

Certainly, therefore, we cannot exonerate the consumer from the environmental impacts of the industrial society. Neither can we expect to clean up our environmental act just by looking at the damaging emissions from the primary and the secondary sector industrial processes. Rather we must place the consumer at the centre of the complex materials network which comprises the industrial economy. And we must place on the consumer at least some of the responsibility for making the economy sustainable.

A COMMON PHYSICAL BASIS FOR MATERIAL SYSTEMS

Despite the complexity of the modern industrial society, its underlying rationale could be said to be the same as the underlying rationale of all societies throughout the ages: to provide for the needs of the men, women and children who constitute that society. Some of these needs are the basic material needs of all biological organisms: food, water and shelter. So we could even argue that part of the rationale for our complex industrial economies is the same as that which governs the behaviour of every other biological species: survival.

It is clear of course that the goals of the industrial society extend considerably beyond mere survival. In poorer countries, survival is often still a luxury. But in the affluent developed nations of the industrial world, people's expectations and goals reach further than mere subsistence. Later in this book, I will return to consider the question of these more complex goals and needs. Here I want to concentrate on the common physical basis which *all* biological species share alike.

So let us start by looking at the physical laws which govern our behaviour in the industrial economy. These laws are the same fundamental laws as those which govern energy and material transformations in all physical systems. In particular, they govern material activities in the ecological systems (or **ecosystems**) which have successfully sustained a wide variety of species for many thousands of years. So whatever the differences which now distinguish human societies from ecosystems, we can certainly learn from the basic ground rules which determine our common physical inheritance.

Two of the most important of these ground rules come from the physical theory known as **thermodynamics** (which means literally 'the theory of the movement or flow of heat'). Each industrial process and each economic activity involves the *transformation* of materials and energy from one form to another. Thermodynamics provides very specific rules and limits which govern these transformations. This theory was developed right at the start of the industrial revolution, primarily to describe the behaviour and optimise the performance of the early heat engines which paved the way for rapid technological progress (see Chapter 2). So it is perhaps particularly appropriate that we should pay careful attention to that theory in the early stages of this investigation.

CONSERVATION LAWS

Amongst the most important of these physical rules are three important **conservation laws.** The first of these is the law of conservation of energy (also known as the first law of thermodynamics). Energy exists in a number of different forms – including gravitational energy, chemical energy, electrical energy, heat, light and motion – all measurable in the same energy units (called joules). These different forms of energy are continually being transformed from one type to another. For instance, the energy of motion is transformed into heat when we apply brakes to a moving car. Chemical energy is transformed to heat through combustion. In fact, it is these energy transformations that allow us to carry out useful work, and provide goods and services in the industrial economy.

The first law of thermodynamics (the law of conservation of energy) tells us that energy is neither created nor destroyed during these transformations. The total energy input always matches the total energy output. For example, when coal is burned, chemical energy is transformed into thermal energy. But the heat output is no more and no less than the energy stored in the chemical bonds of the coal to start with.

A second important conservation law is the law of conservation of mass during material transformations.[5] This law tells us that the total mass of the material inputs to a transformation process is equal to the total mass of the material outputs. It points out that, if we require certain material products, we must find the equivalent material resources to provide for those products. But it also asserts that all of the material resources which we exploit and transform through human activities must end up somewhere – if not in products, then in the environment.

A third conservation law governs the total quantity of each individual atomic element during (non-nuclear)[6] material transformations. This means that the total amount of carbon (for instance) which is released during the combustion of a carbon-based fossil fuel must be the same as the total amount of carbon contained in the fuel to start with. Again this law allows us to count up the environmental burdens associated with the various material transformations which are used to provide services in the industrial economy.

To take an example, suppose that a certain electro-plating facility consumes 1 tonne of the metal cadmium each year in its processes. Suppose further that 900 kilograms of cadmium end up in the metal products. Since cadmium is conserved through the transformation, it follows that exactly 100 kilograms of cadmium are being emitted from the facility in some form as a waste. Even if special filters are fitted to reduce atmospheric emissions and aqueous effluent to a minimum, this conservation law insists that the unused cadmium must be going somewhere – probably to the solid waste stream.

Conservation laws are important to an assessment of the interaction between economy and environment because they insist on rigorous mass and energy balance principles in determining the material throughput associated with industrial activity.

Nevertheless, these conservation laws do not, on their own, provide a complete picture of the interaction between economy and environment. In fact, by themselves, they may even be misleading about the physical constraints imposed on economic activities. Take for example the energy conservation law. If total energy is conserved through all transformations, then why may we not simply go on using and reusing the same energy without ever needing to mine more coal or drill for more oil? Similarly if individual mineral elements are conserved, why should we not use and reuse those minerals over and again for as long as we need to? Why need we ever be concerned about running out of copper or zinc or iron or phosphate? And as for the environmental impacts of the primary sector, why could we not just do away with that sector altogether?

THE SECOND LAW OF THERMODYNAMICS

The answer to these questions is embedded in one of the most famous and most contentious laws in physics: the second law of thermodynamics. This law and its various interpretations are so complex and so far-reaching that it would be impossible to give a detailed description of them all here.[7] The physicist Sir Arthur Eddington described the second law as occupying 'the supreme position among laws of nature';[8] and the economist Nicholas Georgescu-Roegen described it 'as the basis of the economy of life, at all levels'.[9] This law is so crucial to energy and material transformation that we need to have at least some understanding of it, and its implications for the industrial economy.

The first law of thermodynamics talks about the *quantity* of energy during transformation. The second law talks about the *availability* of that energy to perform useful work. The first law tells us that the total quantity of energy after transformation remains the same as the total quantity before transformation. The second law can be interpreted as saying that the same quantity of energy becomes less and less *available* to perform useful work as it passes through successive transformations. The first law talks of the *conservation* of energy. The second law speaks of the *loss* of availability.

One of the first specific formulations of the second law was expressed in the following way: it is impossible to convert a quantity of heat energy into an equivalent quantity of mechanical work. This formulation was part of the early attempts to understand the simple heat engines which were beginning to revolutionise industry. Thermodynamics therefore imposed specific limitations on the theoretical efficiency which could be achieved by industrial processes. These limitations are still relevant to industrial processes today.

A specific example will illustrate some of the implications of the second law. In a conventional thermal power station, heat energy is used to raise steam in a boiler, and the steam is then passed through a turbine which drives an electrical generator. The second law of thermodynamics restricts the efficiency with which heat energy is converted to mechanical work in the turbine. So the electrical energy which comes out of the generator is always less than the heat energy that went in. This is borne out by practical experience. Usually, the efficiency with which heat energy is converted into electricity is in the region of 35–40 per cent.

Given that the total energy must be conserved, this means that around 60 to 65 per cent of the input energy has gone missing somewhere. In practice, we know that this energy emerges from the conversion process in the form of 'low-grade' waste heat. Could we not then use this heat to raise more steam for the turbine? The answer is no. And it is the second law which explains why. The heat energy which comes out of the turbine is at a lower temperature than the heat energy which went in. Because of this lower temperature, it is less available to perform the useful work of raising steam for the turbine. So the second law of thermodynamics is like a law of diminishing natural returns. Energy becomes less and less available to us as it passes through successive transformations.

The second law is in some sense a strange law because it includes considerations which appear to be subjective, such as the 'usefulness' of work. This is one of the reasons why the law has attracted so much philosophical attention. We could say, for instance, that low-grade heat is itself a useful form of energy – particularly for those who live in temperate climes. In fact, one extremely valuable way of improving the overall efficiency with which we consume fossil fuels is to find some useful application of the low-grade heat which is lost from electricity generation. For instance, in some circumstances it could be used to meet domestic space heating requirements. It might also be useful as industrial heat in processes such as drying and baking.

But once we have used this energy for drying (let us say), the second law insists that the energy output from the drying process is even less available to us. And of course this is also borne out by experience: the energy used to drive off water from a substance during the drying process is imparted to molecules of water which are emitted into the atmosphere. As a result of the first law, we know that the total energy of the molecules in the atmosphere must be marginally increased. But this extra energy in the environment is now so dissipated that it is impossible for us to recover it and put it to use.

Energy is conserved, according to the first law. But it gradually becomes more dissipated and less useful according to the second. This is why, in spite of the law of conservation of energy, we must go on mining more coal or seeking new energy resources.

ENTROPY, ORDER AND DISSIPATION

The link between these energetic considerations and the *material* aspects of thermodynamic systems is made through a concept called the **entropy** of the system. Entropy can be thought of as a measure of a certain kind of randomness or disorder in a physical system. More disordered or random states of a physical system correspond to greater entropy. Conversely, the more ordered states of a system correspond to lower entropy.[10]

The entropy of a system is linked to the availability of energy in the system in the following way. The states with greater entropy are those where less energy is available to perform useful work. The states with low entropy have more available energy. It is for this reason that the second law of thermodynamics is sometimes described as the 'law of increasing entropy'. As time goes by – and successive transformations take place – the energy of an isolated system becomes less and less available and the entropy increases. Or in other words, the system becomes increasingly disordered with time, and materials are successively more randomly distributed. These more random distributions of material correspond to states of the system where there is less available energy. For instance, in the case of the energy output from the drying process, the water molecules which carry the (unavailable) output energy are much more randomly distributed than the directed stream of hot air which carries the (available) input energy.

So the general picture presented by the second law suggests that energy and material transformations occur in such a way as to reduce the available energy in the system, and increase the dissipation of materials through the system.

Now we can see why the second law of thermodynamics is such a startling and important piece of physics for the industrial economy. It seems to indicate that economic activity is essentially **dissipative** of both energy and materials. It appears to suggest that things become slowly but inevitably more random, and more disordered, and it becomes less and less possible to carry out useful organisational work.

At first sight, this result seems wrong. How does it tie in with the rapid development of the industrial economy? How can we correlate this result with the vast organisational complexity achieved by the

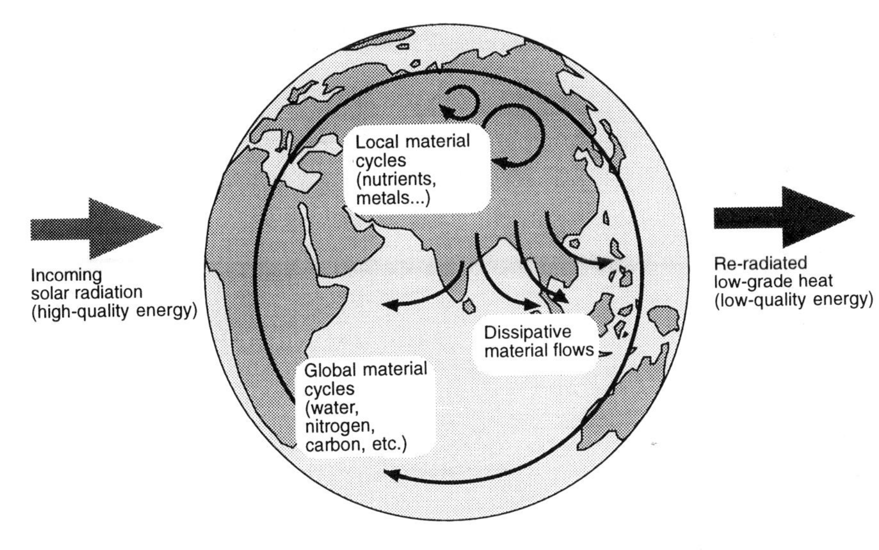

Figure 2 The Earth as a closed[11] thermodynamic system

biological realm? The existence and continuity of life on Earth appear to contradict the doom-laden conclusions of the second law.

This is where the order and disorder interpretation of entropy must be treated with some care. That interpretation was originally formulated by the physicist Boltzmann, who devoted much of his working life to the formulation of statistical thermodynamics. But that interpretation can only be used in a straightforward fashion when it applies to systems which have no external source of energy available to them. Where there *is* such a source of external energy it is possible to maintain order within the system of interest, even though the entropy of the overall system (i.e. the system of interest plus the energy source) must – according to the second law – increase. Fortunately for the process of human evolution, the planet on which we live is in just that situation (Figure 2).

It is the fact that the Earth is subject to a source of solar radiation which allows for the existence of complex biological organisation. Disorder and decay are kept at bay by material transformations made possible by a continuous flow of available energy from the sun. The increase in entropy associated with these transformations (according to the second law) is exported from the system in the form of low-grade heat energy.

Input energy is required in particular to counteract the tendency of materials to dissipate through transformation, and the global ecosystem has developed a complex, interactive network of **material cycles** to accomplish this task. Perhaps the most important of these material cycles is the **carbon cycle**, in which dissipated (i.e. high-entropy) carbon is transformed into fixed (low-entropy) carbon through photosynthesis – the process by which green plants transform solar energy into chemical energy. Photosynthesis forms the basis for a complex food network which supports almost every form of life on earth.

The thermodynamic view of ecosystems which has developed over the last twenty or thirty years[12] is one in which the creation and maintenance of complex organisational structures is not only possible but expected. However, this organisation depends on the existence of well-developed **dissipative structures** which transform high-quality energy inputs to the system into useful organisational work and 'pump out' the disorder associated with that transformation by the second law. The more complex the structure, the greater is the need for high-quality maintenance energy. Complex organisms such as humans require more maintenance energy than simpler ones. Similarly, complex economies require more maintenance energy than simpler ones. In Chapter 9 we shall see that this fact has important implications for the process of economic growth in the industrial economy.

SELF-REGULATION OF ECOSYSTEMS

Using the general outline of material systems provided by the laws of thermodynamics, let us paint a rudimentary picture (Figure 3) of a simple ecosystem, say a small pond, which sustains the life of several species of small fish, numerous plants and insects and a few ducks. What are the physical characteristics of this simple ecosystem? First, there are elements of **self-reliance** within the system. For instance, the fish feed off the plant and insect population in the pond. The waste products of the digestion process (faeces) are excreted into the pond, where they are degraded by bacteria and protozoans (single-celled animals which feed off organic[14] detritus). The products of the degradation process form the nutritional basis for the growth of new plants. This growth only occurs once additional energy is available

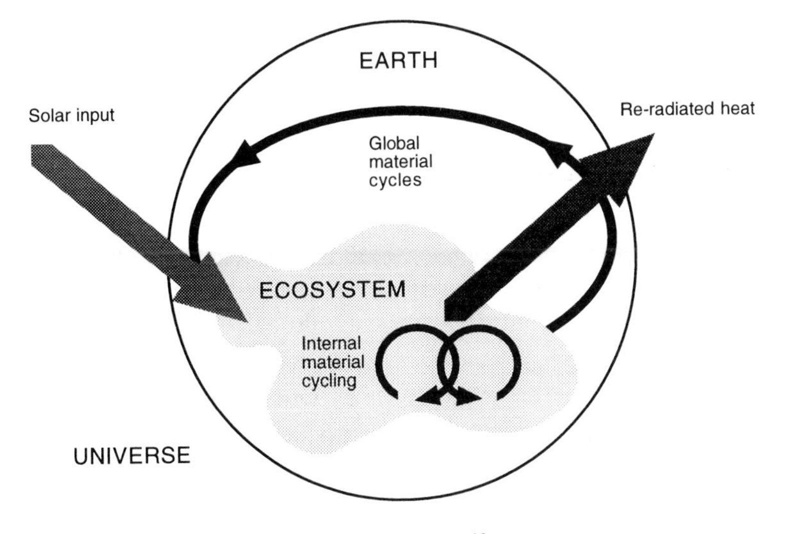

Figure 3 The ecosystem as an open[13] thermodynamic system

from the sun, because high-quality energy is needed to convert the degraded nutrients into the complex molecules needed for biological life.

So the ecosystem recycles some of the materials it needs to sustain life. But it requires high-quality energy from outside the ecosystem to accomplish this task. And it also uses some materials which come from outside the system. For instance, plant life in the pond 'fixes' some carbon from carbon dioxide in the atmosphere. In this sense, each ecosystem is reliant on material cycles which are essentially global in nature, and the picture of the environment as a whole is therefore one in which the interaction between ecosystems is as important as the interactions within each ecosystem.

There are some inherently **self-regulating** properties in the ecosystem. The total rate of activity in the pond is limited by the total available energy flow, which is dictated primarily by the available sunlight, and by the supply of nutrients. Within these constraints it is not possible for the population of any one species to grow beyond a certain size.

Suppose that for some reason there was a small increase in the duck population at the pond. More of the plant and fish life would then be consumed as food, and more faeces would be generated. The increase in organic detritus would fuel increased activity in the protozoan population, leading to an increased supply of degraded nutrients. Provided that there was sufficient incoming radiation to convert the extra degraded nutrients into new plant life (and therefore fish life), the ducks' food supply might continue to match their growth in numbers. But once the limit of the incoming energy had been reached, and no more plant growth was possible, the food supply would run out, and the duck population would gradually decline again until an equilibrium was reached.

This self-regulation is what has allowed many such ecosystems to evolve, to mature and to sustain themselves over periods of time which are staggeringly long in contrast to the speed of change within modern-day human society. It is precisely this self-regulation which is being threatened, in many cases, by the impacts of the industrial economy.

Consider, for instance, the different influences to which our simple pond ecosystem might be subjected if it were situated in the middle of a busy city park. To start with, it will probably be subject to nutrient **run-off** from the fertilisers which have been used in the park to promote the quality of artificial lawns and the flowers. Second, it will be subject to the **fallout** from air pollution: a cocktail of substances, some of which add to the nutrient content of the pond, others which increase the acidity of the water and the mobility of toxic metals in the water, some of which simply deplete the oxygen supply. Finally, the pond may also provide the focus for family outings, where parents and children regularly arrive at weekends to feed bread to the ducks.

Each of these external factors has potentially damaging impacts on the sustainability of the simple ecosystem. Even the apparently charitable attentions of small children can have a disruptive influence on the stability of the simple pond ecosystem. Feeding bread to the ducks means that one species suddenly acquires a preferential source of high-quality energy. This eventually leads to an increase in the duck population. Since their food supply is independent of the natural regulatory mechanisms of the ecosystem, the duck population is no longer limited by the incident sunlight, but instead can grow to a size considerably larger than one which could be sustained by the ecosystem in its natural state. But the results of this unbalanced population growth are potentially catastrophic for the stability of the ecosystem. Increased faeces from the ducks' digestion of bread leads to rapid growth in the bacterial and protozoan population. The excess of degraded nutrients (possibly increased by the run-off from artificial fertilisers) cannot be turned into plant life at the same rate as it is being generated because the available sunshine remains the same as it ever was. The increased protozoan population depletes the oxygen content, inhibiting fish and plant life, and the water becomes clouded with organic detritus.

In the meantime, as their natural source of food within the pond diminishes, the ducks become more and more reliant on their Sunday visitors. When winter comes, and children stay at home, the inflated duck population overgrazes the surrounding grass, which becomes increasingly barren. The park-keepers try and solve the problem by spreading more fertilisers. And so the situation worsens. Finally, all that is left is a murky and virtually lifeless pond surrounded by barren earth around which the unfortunate ducks clamour ungraciously for bounty dispensed at the hands of fickle and unwitting environmental vandals!

THE STRUGGLE FOR EXISTENCE

Of course, I have painted a deliberately pessimistic picture. All the same it is one that is not entirely unrecognisable. Very few ponds in industrial countries still achieve the careful balance of a self-sustaining ecosystem. But this picture of the destruction of natural balances is bleak for a number of reasons. Amongst them is the undeniably good intent of those who feed the ducks. Charity towards those species which share the world with us is a relatively rare quality in twentieth-century industrial society. So it is perverse that it should contribute to – rather than prevent – environmental damage in one of the few places where it remains. Bleaker still – to the Western eye at least – is the vision of nature's harsh equilibrium in the absence of such charity. Within that vision, the fate of the birds – indeed the fate of every living creature – is to struggle for existence in a hostile and competitive natural environment which is strictly limited in its material and energy endowments. Mobility, food, reproduction, defence and shelter are all purchased at a premium – every one of them paid for from a strictly limited supply of available energy. And there is no opting out of it. Even the simplest organisms require a constant supply of such energy simply to maintain life. As Boltzmann himself once remarked: 'The struggle for existence is the struggle for [available] energy.'

I think what this example points out is how difficult environmental management is, and how fragile are the complex materials balances which support survival in the natural world. So I am certainly not recommending we chastise charity towards nature. Rather, I am using this story to illustrate how important it is to develop the complex, comprehensive understanding which a truly charitable attitude to the environment demands.

THE ESCAPE INTO AFFLUENCE?

Human society appears on the surface to have freed itself from nature's limitations more or less completely. This is especially true of the affluent industrial societies in the developed world, in which energy is plentiful, food is abundant, and manual labour has been reduced to a minimum. In fact, much of our industrial creativity in the late twentieth century is dedicated to further relief from the drudgery of household chores and the exploitation of increased leisure time.

This achievement is remarkable in its own right. It is particularly striking, however, when we consider that human society is bound by the same immutable physics that seems to render the conditions of life so hostile for all other species. How can this have happened? Is humanity exempt from the second law of thermodynamics? Or if not, how is it that we have managed to free ourselves from Boltzmann's 'struggle for existence'? The answer to these questions is at once devastatingly simple and emphatically critical of the path of human development.

First of all, we should be quite clear that there is no sense at all in which human activity can escape the constraints of thermodynamics. The material interactions of the industrial economy are subject to the same physical laws as those of the simple ecosystem. In particular, the conservation laws are the same, and the second law of thermodynamics continues to hold. This law is particularly important because it characterises economic activities (like ecological activities) as

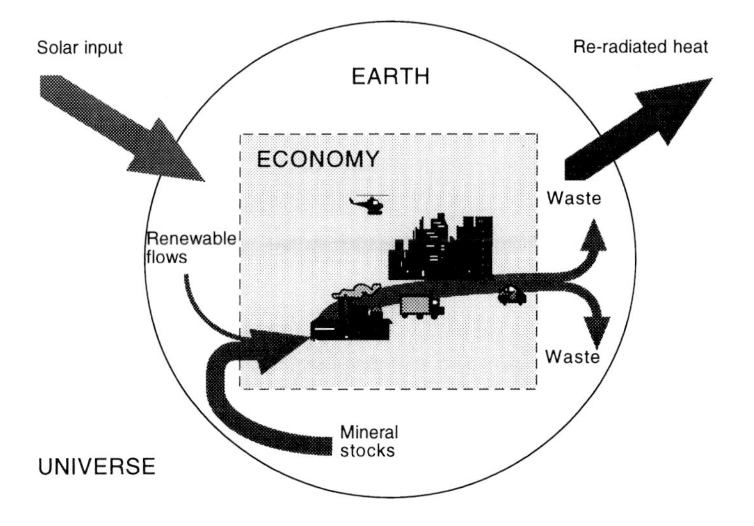

Figure 4 The material economy as an open thermodynamic system

essentially dissipative of both energy and materials. So far so good: the industrial economy behaves exactly like the pond – sustaining life by dissipating energy and materials. But there the analogy ends. The supply of high-quality energy which maintains such a delicate balance in the pond is the solar input. The material inputs and outputs form a part of global material cycles where degraded materials are returned naturally to available states using solar energy. By contrast, the industrial economy has freed itself from the constraints of the solar inheritance by learning to access vast stores of high-quality chemical potential energy locked into mineral resources (Figure 4). In particular, of course this high-quality energy store includes the fossil fuels: coal, oil and gas. These reserves of low entropy provide high-quality thermal energy through combustion. And this energy is then transformed into a variety of other types of useful energy: mobility, light, electricity and mechanical work. Armed with this extra supply of high-quality energy, industrial society has gone one step further and used the additional power available to it to access material resources which are simply unavailable to other species.

These differences are absolutely crucial. The dissipation of materials by the simple ecosystem is limited to the degradation of nutrients and minerals which subsequently return to natural material cycles powered by solar radiation. These constraints provide a kind of natural regulatory mechanism which maintains complex material balances between different species within the ecosystem and between different ecosystems. The economic system provides little or no regulation over material dispersion. It dissipates a wide array of chemicals, some of which do not even exist freely in nature, and many of which exceed the natural flows by several orders of magnitude.

THE DILEMMA OF INDUSTRIAL DEVELOPMENT

These considerations seem to place us on the horns of an uncomfortable dilemma: pursuing our present course appears to offer us a high material standard of living but threatens to destroy the stability of our natural environment; forgoing the benefits of industrialisation might save the environment, but threatens to return us to a savage struggle for survival.

The goal of environmental management in the industrial economy has really been to escape from this dilemma, to find a path for development which retains the advantages to human welfare which have been achieved through industrialisation and yet allows for the future health of the environment. The constraints of thermodynamics suggest that this task is not an easy one. Because all activities are essentially dissipative of both energy and materials, any attempt to clean up environmental damage 'after the fact' is itself a dissipative process, emitting its own inventory of pollutants, and sometimes doing no more than pushing pollution from one place to another.

Preventive environmental management – which provides the focus for much of the rest of this book – searches for a way out of the dilemma described above by redesigning and reorienting economic activities. The object of this reorientation is to try and make the industrial system more like a natural ecosystem: creating material cycles; improving material and energy efficiencies; reducing dissipative consumption; and improving the utilisation of our natural solar inheritance. The following chapters of the book reveal that this task is a challenging one, demanding creativity and innovation at every level: technical, economic, institutional and social. Under the strict constraints of the physical world, however, this search offers us our best – and perhaps our only – hope of developing in harmony, rather than in conflict, with nature. As the economist Georgescu-Roegen remarked: 'In a different way than in the past, man will have to learn that his existence is a free gift of the sun.'

2

MATERIAL TRANSITIONS
The birth of the industrial economy

INTRODUCTION

Sweeping changes have characterised the development of human society over the past 250 years. Revolutions in agriculture, industrial technology, transport systems, and communications have been accompanied by massive changes in the material base of the emerging system. Transformations of the material base have brought changed environmental burdens: a vast increase in the extent and the range of material impacts imposed by human activities on the environment.

In the late twentieth century, we tend to take most of these changes for granted. We generally suppose that industrialisation was and is the most obvious goal of human development. We assume that economic growth achieved through industrialisation is a sign of progress. If asked to define what we mean by progress, we would probably point to rising living standards, improved material comfort, better health, reduced manual labour and increased life expectancy. And generally speaking, we conveniently omit from our deliberations the uncomfortable environmental and social costs which this kind of progress has brought with it. Nor do we usually question whether the same development paradigm is appropriate – or even accessible – for global consumption. We assume that it is.

It is natural enough, of course, to take for granted the basic elements of the culture into which we are born: the philosophies which guide it, the creeds which govern it, and the institutions which embody it. As participants in a particular culture we are often conveniently blind to the fact that culture is relative.

In another world or in another culture, this blindness might not matter. Another culture might automatically have built environmental protection into its lifestyle, for instance, through strict codes of conduct governing material transactions.[1] Equally, in a world without physical constraints, our own lack of constraint might not matter. In a world without the second law of thermodynamics, for instance, materials would never run out, energy would be unlimited, and we could effortlessly contain our material profligacy. Consequently, there would be no material or environmental limit to the global extension of the culture.

Ironically, this is almost precisely how the culture in which we are now embedded has behaved: as if there were no second law of thermodynamics. This shortsightedness becomes more comprehensible when we learn that the second law was not really formulated until Carnot's work in the early nineteenth century. And by that time, the conditions for the development of modern industrial society were already more or less in place. The emergence of the market economy had already occurred. The industrial revolution was in the throes of transforming life in Britain, and was soon to transform life in the other northern European countries and in America.

Since the history of this process is fascinating in its own right, and since a cursory knowledge of it is the least we need to understand our own position in the industrial economy, this chapter is dedicated to a brief history of the material transitions which have taken place in the last 250 years.

A reader who is already familiar with this history or who has absolutely no interest in it could skip to Chapter 3 without losing the main technological thread of the argument. On the other hand, there are aspects of the present chapter – particularly those concerned with the ideological underpinnings of the modern economy – which are crucial to a proper understanding of the latter part of the book.

THE NATURE OF THE INDUSTRIAL REVOLUTION

The industrial economy was born in one particular country – Britain – in the period which is usually referred to as the **industrial revolution**: loosely, between the mid-eighteenth century and the mid-nineteenth century. The time period is approximate. Considerable industrialisation occurred after 1850, and the period before 1750 was characterised by what has been called proto-industrialisation:[2] the development of technical, economic and demographic conditions which prepared the way for industrialisation. Nevertheless, the period between the mid-eighteenth century and the mid-nineteenth century – the period when Britain became the 'workshop of the world' – provided the foundations for one of the most remarkable technical, economic and social transformations in recorded history.

What characterised this transformation? What exactly was the industrial revolution? Historians have tended to accord three slightly different meanings to the term,[3] each of which captures something of the extraordinary metamorphosis that was taking place.

Most narrowly, the term is used to describe the very rapid growth of certain manufacturing sectors in a certain country – primarily the cotton industry in Britain – over a relatively short time interval: between about 1760 and 1840. The technological dimension of this phase of industrialisation was relatively simple. Simple technologies – some of which had been around for a while, and mostly using conventional energy sources and materials – were introduced into the textiles industry. Rapid deployment led to massive increases in output over a very short period of time.

Secondly, it is used to refer to a very specific structural shift which occurred in the British economy over a slightly longer time period. This structural shift was essentially a change from an economy based predominantly on local craftsmanship and agriculture, using simple technologies and renewable resources, to an economy based predominantly on factory-based, manufacturing industry whose raw material basis was increasingly supplied by mineral resources.

Finally, the revolution is sometimes denoted as one in which the entire British economy – and subsequently the economies of Britain's European neighbours – broke away from a system of more or less steady-state national income to a system characterised by continued growth in the national income. This economic revolution was really the emergence of what has become known as **capitalism**: the systematic pursuit of profit through accumulation of private investment capital in a market economy.[4]

The time-scale over which this aspect of the revolution is supposed to have occurred is longer than those of the previous two characterisations. The foundations for the pursuit of private profit were laid well before the middle of the eighteenth century. It was evident, for instance, in Britain's massive expansion in naval power from the middle ages onwards. It was evident in the enclosure movement, in which large areas of land, previously considered common property, were enclosed by private landlords. And it was reflected in an immense body of Western thought stretching (at least) from Adam Smith[5] and John Stuart Mill[6] to Alfred Marshall in the early twentieth century.[7]

Whichever characterisation of the industrial revolution we choose to adopt, the truth of the matter is that all of these changes and more took place during what is a remarkably short period of human history. And in their wake, the simple, material basis of predominantly agricultural human settlement was completely overturned. A few statistics will help to illustrate this startling transition.

THE POPULATION EXPLOSION

The population of Britain had remained relatively stable during the sixteenth and seventeenth centuries, and even during the early part of the eighteenth century. It is estimated to have just about doubled from 3 million to 6 million in the 250 years prior to 1750. But from 1750 onwards it rose dramatically, doubling in the fifty or sixty years following 1780, and doubling again in the sixty years from 1840 to 1900. The population of Britain in the last years of the twentieth century is more than ten times its pre-industrial level. Similar trends occurred in other industrialising nations.

Although population rates are now stabilising in most industrial countries, worldwide population already stands at just under 6 billion and is projected to reach 10 billion people before the middle of the twenty-first century. The cumulative effect of this population increase on the environment is obvious.

One of the most commonly held views about industrialisation is that it improves the quality of life, and increases life expectancy. In reality, little of the early population increase is believed to have been the result of reduced mortality from improved conditions. Basic health services lagged some considerable time behind the increase in industrial output. And in many industrial towns and villages in Britain, living conditions were considerably harsher than they had ever been before.

Life expectancy in Dudley, Worcestershire, in the middle of the nineteenth century, for instance, was just eighteen and a half years.

It is more likely that the early increases in population resulted from a rise in the birth rate. And this birth rate increase was influenced by several factors. Higher incomes meant that more people could afford to have bigger families. But in addition, there was an economic incentive to increase the family size: more children meant more opportunity to take advantage of the increased requirements for industrial labour that the revolution provided. Industry was expanding so fast that its new factories produced a massive demand for labour.

ENERGY AND OUTPUT IN THE 'WORKSHOP OF THE WORLD'

Increases in production output were spearheaded by the textiles industry. Between 1760 and 1787, the output of the cotton industry had increased almost tenfold from just over 1,000 tonnes to around 10,000 tonnes and had jumped to well over 150,000 tonnes by 1840.[8] But the fortunes of the cotton industry quickly affected the rest of the British economy. The application of the steam engine in 1779 to the textiles industry stimulated both the coal industry and the iron industry.

The coal industry was on a relatively stable footing even prior to the industrial revolution. Coal production had increased following drastic fuelwood shortages during the sixteenth and seventeenth centuries, and to meet the demand from Britain's expanding overseas trade. When the population began to escalate in the late eighteenth century, the market expanded further. And once the steam engine and the coking process had revolutionised British industry, the demand for coal soared. From an output of around 3 million tonnes prior to the industrial revolution coal production had increased to 11 million tonnes by the last decade in the eighteenth century and to 36 million tonnes by 1830. At this time Britain more or less completely dominated the world coal market, producing 70 per cent of the total world coal production of around 50 million tonnes.[9]

Cheap transport was one of the reasons for the success of the British coal industry both at home and abroad. International trade was facilitated by Britain's immense naval resources. Domestic trade was also initially dependent on water transport (port-to-port shipping and the extensive canal system). Later, of course, the railways took over from waterways as the dominant transport medium for industrial products. So the development of rail transport both relied on and contributed to the development of the coal industry.

By the end of the ninteenth century British production had reached almost 250 million tonnes of coal, although its domination of the world market (700 million tonnes) had declined. During the twentieth century the dominance of coal as an energy source began to be eroded, first by oil; later, although to a lesser extent, by gas. Domestic consumption of coal in Britain peaked in the mid-1950s at around 220 million tonnes and has since fallen to around 100 million tonnes. But total energy consumption has continued to rise. The UK now consumes an amount of energy equivalent to 350 million tonnes of coal every year. Worldwide annual consumption of primary fuels now stands at around 12.5 billion tonnes of coal equivalent.[10] And this total has more than doubled in the fifty years since the end of the Second World War.

These statistics portray a massive increase in the energy intensity[11] of human society over pre-industrial levels. Even once the increase in population has been taken into account (Figure 5), per capita energy consumption in Britain increased by a factor of about 20 between 1700 and 1990.

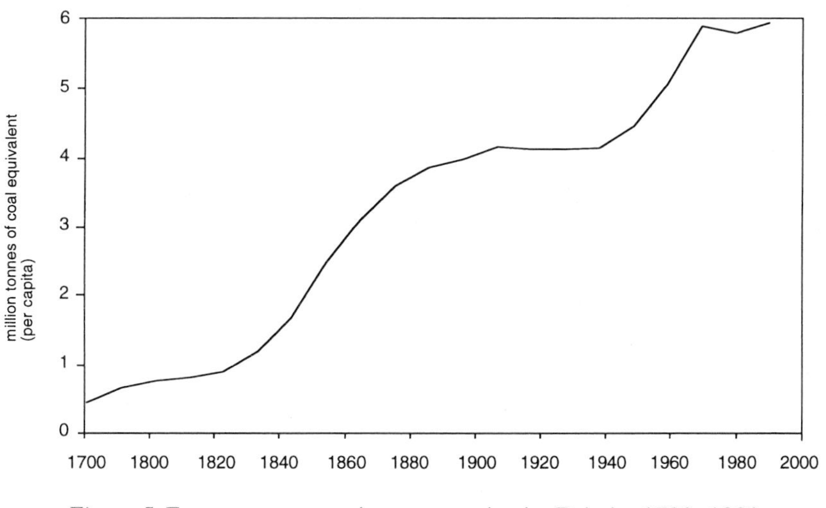

Figure 5 Energy consumption per capita in Britain 1700–1990

ENERGY SURPLUS AND MATERIALS EXPANSION

From a thermodynamic viewpoint, the rapid expansion in the use of fossil fuels has allowed us to escape from the 'struggle for existence' (see Chapter 1) which characterised previous agricultural societies and still characterises the development of almost every other species on the planet.[12] A simple calculation of the energy 'surplus' available through the employment of fossil resources is informative.

In the late eighteenth century a miner could extract around 500 pounds of coal a day, that is, a little under a quarter of a tonne. This quarter tonne of coal could provide around 3,000 times the energy the average miner expended to produce the coal. Even if this thermal energy was converted into useful work with a very limited efficiency, say 1 per cent, it was still 30 times more energy than the miner could provide by carrying out the work of the machines manually. As technical conversion efficiencies improved, this equation swung even more in favour of the new energy system.

It was really this immense new source of available energy which fuelled the continuing process of industrialisation. The early innovations in cotton manufacture were principally simple technologies often driven by water power, rather than coal. But as time went on the energetic advantages of fossil fuels became increasingly important to the industrial economy. Coal provided thermal comfort to the increasing population. But it also provided a more mobile and adaptable source of energy for the vast new factories and industrial towns which were springing up. It facilitated (and also depended on) the massive revolution in transport which industrialisation was spawning. And it provided the energy needed to access an enormous material base of mineral resources: iron initially.

By the end of the nineteenth century the production of pig iron in Britain had increased 200-fold from a pre-industrial level of 50,000 tonnes per year to something approaching 10 million tonnes per year, approximately a quarter of the global production of 40 million tonnes.[13] Today the world consumes over twenty times this amount, and the long-term trend is still increasing.[14] Moreover, the industrialised world now consumes vast quantities of other mineral resources: non-ferrous metals like copper, lead, zinc and aluminium, non-metallic

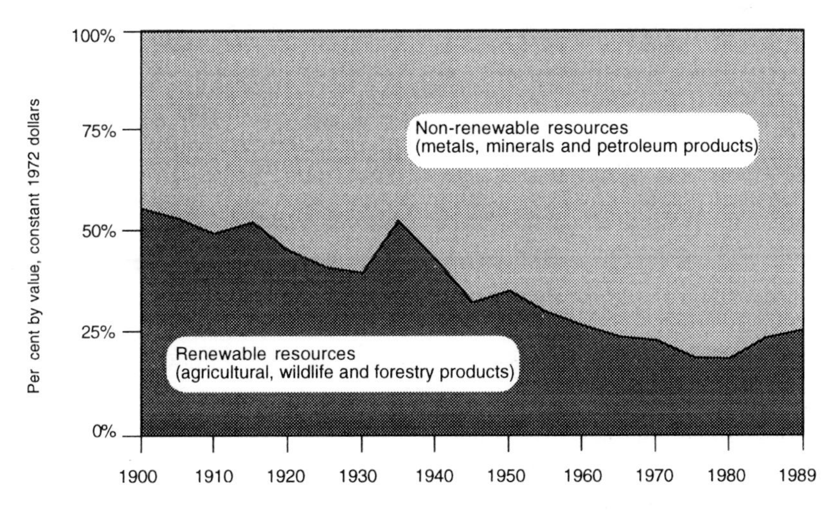

Figure 6 Consumption of raw materials by type in the US (excluding food and fuels): 1900–89

minerals such as stone, clay and sand, and petroleum. Only a few of these minerals have witnessed any decrease in consumption levels over the last twenty years.

In general terms, both the volume and the variety of materials consumed in the industrial economy are still increasing. And the relative changes in the composition of the resource base indicate that, increasingly, these materials are drawn from finite (non-renewable) sources.

Figure 6 illustrates some of the qualitative changes which have taken place in the material basis of the industrial economy since the beginning of the twentieth century. In 1900, even after 150 years of industrialisation, over half of the total materials in use (excluding those used for fuels and for foods) were still provided by agricultural, wildlife and forestry products. By 1989, the proportion of such materials had dropped to less than 30 per cent. Over 70 per cent of the materials basis is now provided by non-renewable resources: metals, minerals and petroleum-based products.

CHEMICAL COMPLEXITY

Some of this change is attributable to the emergence of one particular industry. The chemicals industry is supposed to have been born during the industrial revolution as a result of a suggestion made by Berthelot to James Watt in 1786 that chlorine could be used for bleaching. But the emergence of the chemical industry might equally well be attributed to the demand for soda ash – used mainly to make glass, soaps and cleaning products. The Leblanc process for soda ash production was patented in 1773, a quarter of a century before the first chlorine patent (for the Deacon process in 1799). At any rate a flourishing chemicals industry had emerged in Scotland by 1800, and has expanded almost continually ever since. The most dramatic impacts have been seen in this century, particularly since the Second World War.

The chemicals industry in the United States, for example, has expanded more than tenfold in the last forty years. Approaching 100,000 industrial chemicals are now in commercial use worldwide, and this figure is increasing at the rate of between 500 and 1,000 new chemicals each year. This increase has been driven in part by the availability of petroleum-derived by-products of an expanding oil industry, and in part by the increased role for new and complex chemicals in new and expanding technological contexts: agriculture, metal purification and metal plating, electronics, textiles and the food industry.

Some of these chemicals are known to be toxic to humans. But scientific information about the toxicological and environmental impacts of many other chemicals is simply inadequate. The US National Academy of Science has estimated that there is insufficient information even for a partial health assessment for roughly 90 per cent of them.

A global assessment of the total toxic emissions into the environment from manufacturing industry (Figure 7) reveals that almost 40 per cent of this burden comes directly from the chemicals industry – the single biggest source of toxic emissions.[15] The second biggest category is the metals industry, accounting for 26 per cent of total emissions. But the burden of toxic releases from other industrial sectors should also be accounted as indirect effects of the chemicals industry. Many of its toxic products are destined for consumption in other industrial sectors, and a large proportion of them will be emitted elsewhere in the economy, either as process emissions from other manufacturing sectors, or through product use and disposal.

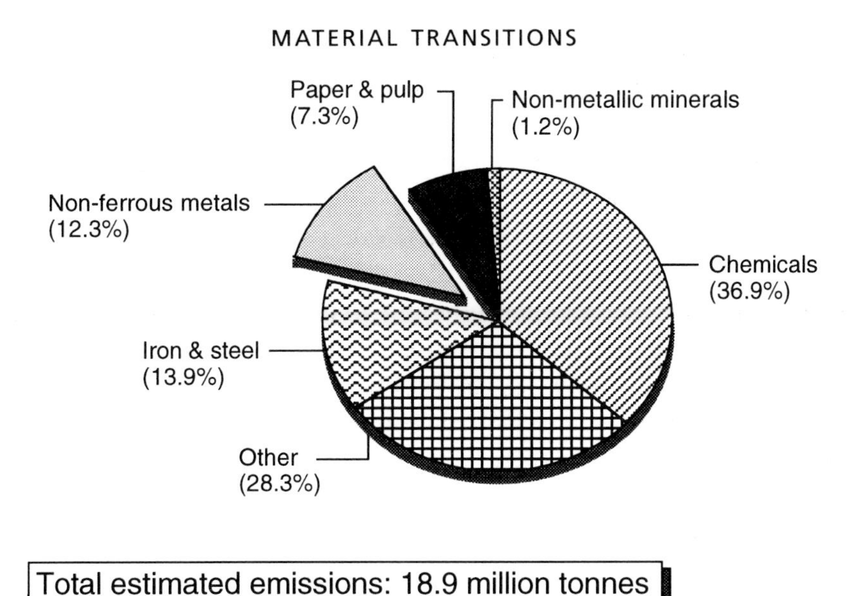

Figure 7 Global toxic emissions from manufacturing by sector (1990)

The new materials which are now being developed are no less problematic. For instance, the new generation of materials used for semi-conductors and superconductors increasingly uses rare metals, on which very little research on health effects or environmental impacts exists. New composite materials which combine polymers with glass or metal fibres offer high occupational hazards during manufacture and pose serious waste problems after use. Perhaps the most worrying concerns come from new developments in the biological sciences. These developments in what is now known as **biotechnology** present many of the most troubling characteristics of environmental concern: novel and inherently uncertain ways of manipulating biological resources; proposed application in an open and complex environmental system; and an inherent lack of control over released materials and processes.

THE SHADOW OF ENVIRONMENTAL POLLUTION

The environmental impacts of the new industrial age were felt within a relatively short space of time after the industrial revolution, and have gradually gathered an insidious momentum. Early problems were mainly confined to air pollution from the burning of coal and the smelting of ores, and water pollution from foundries and metal-works. In addition, the social restructuring and urbanisation of the industrial society gave rise to some local sewage-related problems.

The twentieth century has seen a ballooning of other, less manageable forms of pollution: the emission of greenhouse gases (such as methane and carbon dioxide) which threaten to warm up the global atmosphere; the emission of chlorofluorocarbons (CFCs) and other halogenated[16] chemicals which threaten to destroy the ozone layer; the release of toxic, synthetic chemicals such as dioxins, and polychlorinated biphenyls (PCBs) which are harmful to human health in very small quantities; the emission of gases like sulphur dioxide and nitrogen dioxide which combine with water in the atmosphere and lead to the acidification of soil and water supplies; the generation of hazardous and toxic industrial wastes which are emitted into the atmosphere, or discharged into rivers and lakes; the accumulation of industrial and household wastes in landfill sites designed for the disposal of products; and a sophisticated cocktail of transport-related pollutants giving rise to dangerous photochemical smogs in large cities. There is now plenty of literature which deals in detail with these various forms of pollution from the industrial economy,[17] so I do not intend to dwell on them here. Rather I want to stress again the link between environmental pollution and the dissipative nature of economic activity. Without being simplistic, we could say that almost all the environmental problems of the industrial economy arise from the dissipation of materials through the economic system.

We know from the second law of thermodynamics (described in the previous chapter) that this dissipation is inevitable. The same inevitability holds for natural ecosystems. The difference between the economic system and the ecosystem is that material dissipation in the economy is independent of the complex balance of natural material cycles which reorganise degraded materials into high-quality resources again. And the dissipation of energy is largely independent of the powerful solar flux which preserves order in the global ecosystem. By contrast, the dissipation of materials by the industrial system proceeds without inherent limitations or regulations. Accessing huge reserves of available energy has propelled human civilisation into a new industrial era. But environmental pollution has followed that trajectory, as the inevitable shadow of our material progress.

SOCIAL IMPACTS OF INDUSTRIALISATION

These remarks serve to illustrate why the technological and material transitions which have taken place since the industrial revolution carry such profound environmental implications. But whatever the industrial revolution was, it was certainly something more than simply technological change. And although rapid demographic changes occurred over a period which seems perilously short by geological timescales, these *quantitative* changes were matched by equally startling *qualitative* changes in the way in which people lived and made their living.

In 1750, even though the wealth of the country was certainly no longer tied to agriculture, the population of Britain was still predominantly rural. Perhaps 80 per cent of its population lived in the country.[18] London had a population approaching three-quarters of a million people, which was enormous by the standards of the day and twice the size of its nearest rival, Paris. But no other town had a population greater than 50,000 inhabitants. Even in 1800, with the industrial revolution well advanced, the rural population comprised about 70 per cent of the total. By 1950, the rural population had fallen to less than 20 per cent of the total. And rural populations have continued to decline.

This change was a direct consequence of the industrial revolution. The needs of the new mechanised industries were for cheap, plentiful labour, capable of being organised in such a way as to maximise the returns on capital investment. This inevitably led to the emergence of new industrial towns centred around the new factories where human labour could be concentrated and organised. Above almost every other characterisation of the industrial revolution rises the image of the new urban factories. To some this image was a positive vision, promising a new prosperity and providing a new, vibrant social environment. The reality was less than utopian. Poorly paid, overworked labourers, often women and children, eked out a meagre existence in appalling working conditions. In those early days, environmental management did not even extend to humane working conditions.

One of the criticisms levelled most often against this new form of social organisation was that it destroyed the independence and the creativity of the rural artisans: spinners, weavers, potters, carpenters, ironmongers and tanners. It was in fact this flourishing artisanal community which had provided the domestic basis for the early period of commercial proto-industrialisation. Contemporary accounts of Britain in the early eighteenth century portray the rural villages as prosperous, well-organised environments, flourishing in the improved commercial situation. Ironically, it was precisely the success of this new commercial basis which was eventually to rob the artisans of their employment.

In any case, the emerging reorganisation of labour certainly created a completely new social environment for the vast majority of the population. People flocked to the new industrial towns and cities in their thousands. Agriculture benefited from the new industrialisation. Better ploughshares, improved transport, new energy sources for irrigation, and new crop rotation techniques: all these contributed initially to significant improvements in agricultural productivity. And these improvements themselves contributed to industrialisation by freeing labour from farms for the new industrial factories.

The urbanisation of populations and the centralisation of production created new infrastructures with new material demands. Different institutional structures emerged. Different economies of scale applied. Amongst the most fundamental changes were those which applied to transport. Transport systems underwent several radical revolutions as they strove to meet the demands of centralised production and concentrated populations. But the new transport systems had their own impacts on the environment. The automobile may have become a symbol of personal freedom in the late twentieth century, almost in itself an icon of modern society. But it is also one of the biggest sources of environmental concern in almost every industrialised nation in the world.

Equally importantly, the new urban demography had some profound and far-reaching impacts on the social behaviour of the industrial economy. People were no longer in daily contact with their natural environment. They no longer perceived the importance of the changing of the seasons, or the climate, or the quality of soils and rivers. They became gradually more and more isolated from the source of their own survival. And this in its turn allowed a social evolution almost entirely divorced from any awareness of natural, physical constraints. Even television, the omnipresent provider of diffuse information in the late twentieth century, has been unable to replace our close, experiential knowledge of nature. Today, barely a fraction of the population possesses skills relevant to their own survival as animals in a natural environment. And only a few more have anything other than a cursory understanding of environmental constraints.

These social aspects of industrialisation are important to our investigation for several reasons. Social repercussions contribute to the environmental burden in various material ways. Social trends have subtracted from our environmental proficiency, even in its most basic sense. And finally, we have to be able to understand the driving forces behind the industrial economy if we are to address its environmental problems. If we are to hope to reorient its progress towards a sustainable future, we need to get inside the industrial machine and uncover the driving mechanism, the principles and philosophies which have forged the new material society.

ECONOMIC FOUNDATIONS OF THE INDUSTRIAL ECONOMY

Above all else these principles and philosophies were those encapsulated in the new economics: the pursuit of profit, the accumulation of wealth, the employment of capital resources to improve labour productivity, and the transition from a steady-state economy to one predicated on continued growth.

We can trace these ideas back to well before the industrial revolution. Many authors point to a poem entitled *The Fable of the Bees*, published in 1705 by a Dutch doctor living in London, Bernard de Mandeville, in which he compared the social behaviour of the emerging market economy to the operation of a beehive:

> Vast numbers thronged the fruitful hive;
> Yet those vast numbers made 'em thrive;
> Millions endeavouring to supply
> Each others' lust and vanity. . . .
> Thus every part was full of vice
> Yet the whole mass a paradise.[19]

Essentially, the poem was a satirical attack on the moralists who deplored what they saw as the rise of greed and vanity in late seventeenth-century England. De Mandeville's thesis was that it was precisely these vices of self-interest which ensured progress, and supported civilisation.

Ironically, it was Adam Smith – the 'father of economics' – who first protested against this characterisation of self-interest as the motor of progress. In *The Theory of Moral Sentiments* published in 1759, he described de Mandeville's thesis as 'holy [sic] pernicious' and 'in almost every respect erroneous'. This response is most notable for its appearance in the same book in which Smith first proposed his famous doctrine of the 'invisible hand'. In this early work, however, the invisible hand is synonymous with a kind of divine providence, a guiding force moderating the 'vain and insatiable desires of the rich'.

By the time, he wrote *The Wealth of Nations* in 1776, Smith had quietly succumbed to the logic of de Mandeville's satire. In the later work, Smith argued that each economic agent is led

> by an invisible hand to promote an end which was no part of his intention. Nor is it always worst for society that it was not part of it. By pursuing his own interest he frequently promotes that of the society more effectually than when he really intends to promote it.

This doctrine has been handed down to us today in the familiar guise of market economics: the self-interested, rational economic actor pursuing his own profit motive in an open market offers the best way of allocating economic goods; government is necessary only to ensure that the market is working efficiently.

Irrespective of this intellectual underpinning, there is good evidence that the new economic system was operating well before the industrial revolution. The system of enclosures, under which common land became increasingly concentrated in the hands of rich and powerful landlords, was in evidence from the early eighteenth century. England's prosperous, proto-industrial society was already operating on the foundations of a commercial market economy. Britain's extensive overseas power provided the expanding demand base on which industrialisation itself was built. The foundations for a global dependency on economic growth as the basis for prosperity were already laid.

There was opposition of course. The Luddites were the best known opponents of the new industrialisation. They deplored the destruction of local workmanship and the displacement of labour which the guilds had striven for so long to protect. The fiercest criticism was reserved for the impacts of mechanisation. This is reflected in the commentary of William Cobbett after a visit to Withington in the Cotswolds in 1826.

> Here in this once populous village you see all the indubitable marks of most melancholy decay. . . . A part, and perhaps a considerable part, of the decay and misery of this place is owing to the use of machinery and to the monopolizing in the manufacture of blankets of which fabric the town of Witney was the centre and from which town the wool used to be sent round to and the yarn, or warp, back from, all these Cotswold villages and quite a part of Wiltshire. This work is all now gone. . . . There were, only a few years ago, above thirty blanket manufacturers at Witney: twenty-five of these have been swallowed up by the five that now have all the manufacture in their hands. And all this has been done by that system of fictitious money which has conveyed property from the hands of the many into the hands of the few.[20]

But industrialisation proceeded in spite of criticism. Its foundations were in the event far too strong to be swayed by local criticism, and even the trenchant, intellectual critiques of economists such as Karl Marx and Jean-Charles Sismondi failed to modify the continued pursuit of profit in a capitalist economy. Appalled by the effect which industrialisation was having on the English way of life in the early nineteenth century, Sismondi rejected the idea that the 'invisible hand' could be relied on to provide the best for people and called on the government to intervene. He advocated a return to independent, local production and agriculture.

But at the end of the day, the dislocations of the labour pool were localised effects. In general terms, employment opportunities in the UK increased for at least a century, in spite of mechanisation. And by then, the new economic and social system was completely entrenched. Mechanisation increased output, which increased profit, and allowed for increased investment. Today, the pursuit of economic growth is the goal of almost every government in the world.

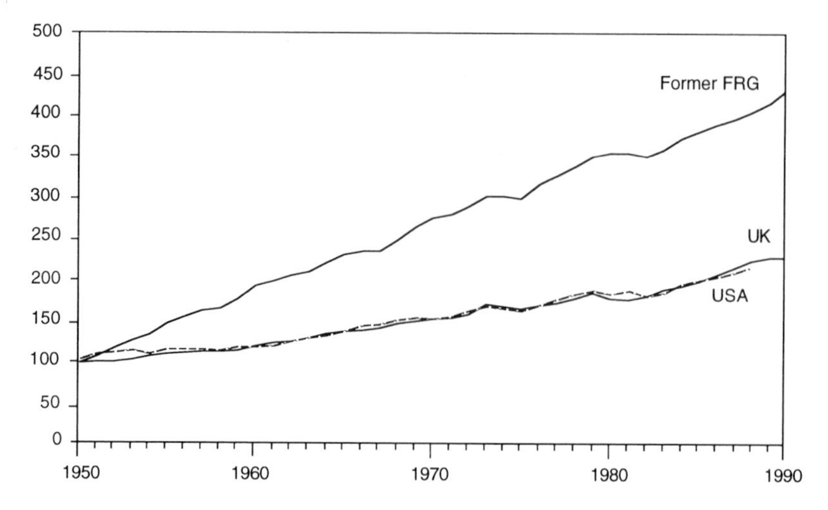

Figure 8 Growth in GNP in three industrialised nations 1950–90 (GNP is set at 100 in 1950 and subsequent years are indexed relative to the base year.)

Figure 8 shows the increase in gross national product (GNP) in three industrialised nations over a forty-year period between 1950 and 1990. GNP is a measure both of national production output and, simultaneously, of national consumption, and so provides a proxy for economic development. The pursuit of rising GNP is now common to almost every nation in the world, developed or developing. More than 200 years after the birth of the industrial age, GNP in Britain still increased by 230 per cent over a forty-year period.[21]

SUMMARY

In the space of a couple of centuries, industrialisation has revolutionised not only the material basis of human society but its demographic underpinnings, its economic structure, its cultural inheritance, its social patterns, and its knowledge base.

For the purposes of the investigation in this book, it is the material transition which is of principal concern. And this transition has been absolutely fundamental. From a society dependent mostly on local supplies of natural, renewable resources, we have become a society dependent increasingly on global patterns of consumption based on limited resources, and the utilisation of exotic, often toxic materials, many of which were completely unknown in the natural environment less than a century ago. In this context, the task of environmental management is not just vital to our survival. It is also frighteningly complex. Nevertheless, it is this task which we must now face.

3

FAREWELL TO LOVE CANAL

From industrial afterthought to environmental foresight

INTRODUCTION

From the standpoint of the end of the twentieth century, we could say that environmental management has emerged as one of the highest priorities for industrial development. In practice, it has – until recently – been more or less an afterthought. Nowadays it seems perfectly clear that industrial development must strive for a high degree of environmental foresight. But in the past, this has simply not happened. There are a number of reasons for this. Some of them are institutional; others are economic. Some are the result of poor or inadequate science; others again are really a manifestation of the sheer complexity of the problem.

A major change of attitude towards environmental management has occurred in the course of the last two decades or so. This shift – towards preventive environmental management – provides the motivation for much of the rest of this book. Understanding this change requires some discussion of the context in which it has arisen. So this chapter is devoted to a brief description of some of the principal developments in environmental management which have occurred during the last century. It attempts to appreciate the difficulties – rather than simply condemn the deficiencies – of the past, and to identify clear lessons for the future of environmental management.

INCIDENT AT LOVE CANAL

The story of Love Canal[1] is a monument to the failure of early environmental management practices in the industrial economy. The canal was a small artificial waterway which once flowed into the Niagara River in New York State. It was originally excavated in the late ninteenth century as part of a hydroelectric project. From the early 1940s until the early 1950s, the canal was used by Hooker Chemicals and Plastics Corporation as a dump for its chemical wastes. During that time more than 40,000 metric tonnes of toxic chemical wastes (including dioxins, lindane and arsenic trichloride) were dumped into the conduit. In 1945, an analyst for Hooker predicted that 'the quagmire at Love Canal will be a potential source of lawsuits'. Despite this recognition, or perhaps because of it, the site was filled in by the company, and sold to the local Board of Education in 1953 for $1, on the condition that the company was absolved from any future liability for the site.

An elementary school was erected at Love Canal, and later a housing estate was built. Twenty years later the site was found to be leaking. There were reports among the local population of skin irritations and respiratory problems, and in 1977 tests revealed that the soil and water in the vicinity of the former dump were heavily contaminated with a wide range of toxic chemicals, many of them carcinogenic. The local authorities refused to act at first. But in April 1978, after persistent lobbying from the local citizens' group, a state of emergency was declared by President Carter's government. It was the first time that a national emergency had been declared anywhere as a result of chemical pollution.

The New York State Commissioner for Health ordered the evacuation of 240 families and the dump was cordoned off as a Federal Disaster Area. Lawsuits of $1,400 million have been filed against Hooker Chemicals and Plastics Corporation. But half a century after the damage was inflicted the site at Love Canal has still not been cleaned up.

If Love Canal were an isolated instance in the history of the industrial economy, perhaps it could be dismissed as the irresponsible aberration of a careless corporation. But it is not an isolated incident. A second Hooker dump at Bloody Run Creek, situated just across the road from a water treatment plant serving 100,000 people, has also been found to be leaking. And there are 212 other dumps in the Niagara Falls area alone, containing an estimated 8 million tonnes of hazardous wastes. In fact, there are around 15,000 uncontrolled hazardous waste landfills and 80,000 contaminated lagoons in the

United States. A national priority list of around 2,000 sites has been drawn up under the US Superfund programme, which has the responsibility of making financial provision for cleaning up past contamination. The cost of cleaning up only these national priority sites has been estimated at approaching $200 billion.[2]

The incident at Love Canal is only one among hundreds of thousands of environmental tragedies worldwide. Some of the worst of these tragedies have taken place in Eastern and Central Europe. Environmental management under the former Communist regimes has lagged significantly behind efforts in the Western world to improve the situation, and there are now areas of Eastern Europe where the soils are too contaminated to sustain agricultural produce safe for human consumption. Water supplies are at constant risk of contamination from leaking landfill sites.

EARLY ENVIRONMENTAL ATTITUDES

This litany of environmental mismanagement suggests that we must regard Love Canal as symptomatic of a particular attitude or 'world view' which has typified the institutions characterising the industrial economy: companies, regulatory authorities, and governments. This world view – which has prevailed since the industrial revolution – supported a kind of *laissez-faire* philosophy towards the environment which has allowed waste materials to flow more or less unhindered out of the economy into the atmosphere, into lakes, rivers and seas, and on to the land.

If we were to search for some kind of logic behind this *laissez-faire* philosophy, we would find several influential factors. Some justification for this world view actually proceeds from a limited consideration of the natural world, for instance, from the behaviour of materials in natural ecosystems. Equally influential are the economic and institutional structures which characterise industrial development, and provide powerful incentives for particular kinds of behaviour. But there are also a number of circumstantial features of the problem, factors which perhaps justified a particular world view prior to industrialisation but render it suspect later on.

Amongst the circumstantial factors, for instance, is the obvious element of scale. When the scale of human activities was more limited, so were its impacts. When land was abundant and human settlement small by comparison, there was always room to move away from adverse environmental effects, and to start again somewhere else. In fact, history is littered with instances of environmental migration. For example, the once coastal city of Ephesus in Turkey (which is now several miles from the sea) was moved several times and finally abandoned in the fourth century after siltation rendered the port unusable. Debeir *et al.* (1991) describe the prevalence of this **foul-and-flee** style of environmental management in the pre-industrial energy system in China.

Another circumstantial factor is the changing material nature of human activities. Although the material basis of society is now phenomenally complex, it was once relatively simple. Even as little as fifty years ago, remarkably few synthetic toxic materials were in use. What are now clearly bankrupt environmental attitudes are sometimes no more than an extension of philosophies that might once have been valid. And of course we must also weigh up the new demands on our environmental knowledge which the variety and complexity of the industrial world imposes. Certainly, our existing knowledge is too limited to be able to predict accurately the environmental effects of all these new activities. It might even be argued that the increase in complexity of human activities has introduced new, irreducible uncertainties into the problem of environmental management.[3]

The supporting beliefs – beliefs which to some extent justified the *laissez-faire* approach – are also interesting to contemplate. These beliefs formed a part of the basis for later developments in environmental management. And although the newer developments have not always been successful, they certainly offered short-term environmental improvements over earlier attitudes.

I highlight two particular supporting beliefs. First, there is a belief that it is possible to **concentrate and contain** environmental contamination: to keep it in one place and prevent it from leaking out into the world at large, or leading to human exposure. Second, and to some extent conversely, there has been a belief that it is possible to **dilute and disperse** pollution through the environment to such an extent that it no longer becomes a threat. Usually, the concentrate-and-contain philosophy has been adopted for disposal of wastes on land. And the dilute-and-disperse philosophy has been used to justify disposal into the atmosphere and into rivers, lakes and seas.

Both of these underlying beliefs have some kind of basis in reality. We know very well that particular substances have remained in more or less the same place in the environment for thousands of years. Deposits of metallic minerals, for instance, have remained locked away in rocks and geological substrata until the advent of modern mining techniques. Equally, of course, the dilute-and-disperse philosophy draws some kind of inspiration from the natural ecological mechanism described in the previous chapter, in which materials are dissipated through natural ecosystems and returned to natural 'anti-entropic' material cycles.

The danger behind both of these assumptions is really the same: neither of these patterns of material behaviour – dilution on the one hand and containment on the other – really reflects the long-term behaviour of materials in the Earth as a thermodynamic system. Special conditions can lead to the accumulation of some minerals in certain deposits and keep other minerals hidden away for long periods of geological history. But there is absolutely no guarantee that we can mimic those special conditions. It has often proved extremely difficult to isolate or contain toxic pollutants even over relatively small timescales. That was amply demonstrated by the incident at Love Canal. Improving the isolation, fortifying landfills, creating special containment sites might help in the short run. But the truth is that we have absolutely no guarantee of the success of such a strategy in the long term. We just do not know how a sophisticated cocktail of exotic toxic materials is going to behave over the length of time we would want ourselves and our descendants to be protected from it.

The dilute-and-disperse approach is particularly suspect. We saw in Chapter 1 that materials do tend to disperse in an isolated system. But we have also seen that, in a system open to the input of high-quality energy, accumulation and cycling of materials are the long-term trends. And it is these anti-entropic material cycles which should warn us against a simplistic notion of dilution and dispersal as the basis for environmental management. Materials may disperse during certain periods of a cycle. But transport, sedimentation and accumulation are equally important features of those cycles. In environmental terms, as in life, it is often true that 'what goes around, comes around'!

DISASTER AT MINAMATA

In spite of these contra-indications, the dilute-and-disperse approach has been applied widely within environmental management. It has been assumed that even highly toxic materials can be released safely into the environment, provided that the receiving medium (the water, air or land) is large enough. A notion has been developed that the environment possesses a certain **assimilative capacity**, a level of tolerance to material inputs, below which no harmful effects would occur. This notion has been used to justify emissions of a wide range of pollutants, particularly into the marine environment. Mercury, lead, cadmium, radioactive waste, sewage, and even synthetic organic materials have been dumped into rivers, lakes and oceans on the assumption that they will dilute and disperse. Unfortunately, this approach is fundamentally flawed and has resulted in some spectacular failures. One of the most tragic of these involved the release of the toxic metal mercury into the sea in the 1950s.

Mercury had been discharged into Minamata Bay in Japan from the Chisso chemical factory for fifteen years before the first cases of 'Minamata disease' were recognised in 1953. The assumption had been that the mercury content of the wastes would be dispersed in the volume of water and diluted to a level at which it would present no threat to the human population.

Two unforeseen factors provoked a catastrophe. First, rather than dispersing, the mercury accumulated through the food chain. Small organisms absorbed the mercury, which was then passed on to the larger organisms which feed on them. The higher up the food chain, the higher the concentration of mercury. Eventually this **bio-accumulation** led to very high levels of mercury in the local fish – the staple diet of a certain section of the local population. These people were therefore at great risk from any contamination, and in the event, they bore the brunt of the fatalities.

The second unforeseen element concerned the chemical form of the mercury. Organic[4] mercury is considerably more toxic to living organisms than inorganic mercury because it is more easily absorbed. Most of the mercury in the emissions to Minamata Bay were in the form of inorganic mercury, giving further grounds for believing the disposal practice was safe. In fact, what happened was that inorganic mercury from the effluent was converted to organic methyl mercury by the action of microbes in the sea sediments. This conversion considerably increased the toxicological risk to the human population via the food chain. Between them these two factors spelt disaster for the residents of Minamata Bay. By 1983, deaths from the incident had reached 300, and at least 1,500 people were officially recognised to be suffering from Minamata disease – although 6,000 claimed to have been affected.

THE FAILURE OF DILUTE AND DISPERSE

This example illustrates just how dangerous assumptions about safe levels of release into the environment can be. Our best knowledge of the behaviour of materials released into the environment is extremely sketchy. Perhaps the most extensive knowledge we have about global material behaviour concerns the major nutrient cycles such as the carbon cycle. This is the cycle which is so critical to the vexing question of global warming. But even in this area, the answers to crucial questions are haunted by uncertainty. How much carbon can we release into the atmosphere without affecting the global temperature? Will some of the extra carbon be absorbed by vegetation or the oceans? What feedback effects will increase or decrease the warming effect of increased levels of atmospheric carbon? The accepted wisdom in these issues still holds that there are major uncertainties associated with all of these questions.

Much more obvious dangers are associated with the release into the environment of materials which do not occur freely in nature. In this case there are no well-established material cycles at all, so that any link with the natural materials cycling of ecosystems is extremely tenuous. There are now so many synthetic chemicals that we do not even have a complete grasp of their toxicity, let alone their behaviour in the environment.[5] We do know, however, that some extremely serious environmental problems have occurred as a result of releasing synthetic substances into the environment. For example, the use of the chemical DDT as a pesticide has raised so many concerns about human and environmental health that Western nations have now banned its use in agriculture.

In spite of these failures, a dilute-and-disperse approach played a rather central role in responses to a wave of environmental manage-

Plate 1 'Dilute and disperse' or 'foul and flee'? Atmospheric emissions from industry in Wales
Source: © The Environmental Picture Library/David Hoffman

ment problems which emerged during the second half of the twentieth century. For instance, clean air legislation introduced in the 1950s in response to heavy city smogs in many industrialised countries was largely based on this philosophy. This legislation did take the unprecedented step of setting aside certain areas as 'smokeless zones'. But the main plank of the clean air legislation was to regulate the height of industrial chimneys, under the belief that releasing pollutants higher into the atmosphere would reduce the environmental impacts by diluting and dispersing them (Plate 1).

Even this simple expedient of dispersing atmospheric pollution through tall chimney stacks has proved ineffectual. Under adverse meteorological circumstances – such as prolonged high pressure and temperature inversion episodes – the pollution becomes trapped for days at a time within a local region, leading to severe respiratory problems for the local population. Even if the strategy is successful in carrying pollutants away from the immediate vicinity of the chimney, they may still eventually cause damage. That damage may be far away, even overseas. But this is not a solution. And sometimes the final environmental burden can be greater, if the pollution happens to be transported to very sensitive ecosystems. Acid pollution from sulphur dioxide emissions provides a clear example of this. Some very sensitive woodland areas such as those in northern Scandinavia are now threatened by acid pollution which has travelled hundreds of miles from Central and Eastern Europe.[6]

We can see from these examples that the concept of dilute and disperse is an extremely problematic one. Even though it may take some of its rationale from the behaviour of some materials in certain ecosystems, there is often very little justification for the assumption that material releases from the economic system will be absorbed within such cycles. Belief in the assimilative capacity of the environment often amounts to little more than blind faith that nature will deal with our waste disposal problems for us.

THE 'END-OF-PIPE' APPROACH

A more sophisticated attempt to deal with environmental pollution emerged to some extent in parallel with the formal adoption of dilute-and-disperse strategies. If there were 'safe' levels of emission into the environment, then there might also be occasions on which emission levels exceeded those considered safe. In these circumstances it was necessary to devise technological strategies to stop emissions leaving industrial factories and entering particular environmental media.

This strategy has been dubbed **end-of-pipe abatement** because it relies basically on placing filters, scrubbers, separators and purification plants at the end of emission pipelines. These filters and scrubbers are added on to the end of the emission pipe of a particular industrial process, in order to stop the release of a particular contaminant into the local water or the atmosphere.

Without a doubt, end-of-pipe techniques have been effective in reducing certain pollution problems, particularly those which arise from **point-source pollution**, i.e. emissions from industrial smoke-stacks and pipelines. For example, sewage treatment plants have been used extensively in the Western world to reduce the emission of raw sewage into rivers, lakes and coastal waters. Filters of various kinds have been used to stop certain toxic metals from reaching the local environment. And the addition of scrubbers and filters has helped to reduce the emission of some dangerous air pollutants.

In spite of these successes, the end-of-pipe strategy has some serious drawbacks. First, it is clear that each end-of-pipe technology is a further process to be added on to the existing process. As such, it is subject both to the laws of thermodynamics and to the laws of economics. This means, amongst other things, that it will always be an essentially dissipative process itself, requiring high-quality energy inputs and resulting in low-quality energy outputs. Some of these dissipative flows will in themselves create environmental problems, so that we may be solving one environmental problem only at the expense of introducing another one.

The problem of controlling acid pollution from fossil fuel combustion provides a graphic illustration of these difficulties. When coal and heavy oils are burned, the sulphur content in the fuels combines with air to form sulphur dioxide. This gas is then emitted from the smoke-stack. In the atmosphere, it combines with water vapour to make an acid. This is one of the causes of the well-known problem of acid rain.

The end-of-pipe strategy for controlling acid rain involves fitting **flue gas desulphurisation** units on to power station and factory chimneys. Making flue gas desulphurisation units involves using additional raw materials and energy. More importantly, the main commercial route for flue gas desulphurisation is a chemical process which uses wet limestone to convert the flue gas sulphur into gypsum.[7] Providing sufficient limestone to carry out flue gas desulphurisation even for one large power station means quarrying and transporting large quantities of limestone.[8] And the gypsum by-product presents a considerable waste disposal problem.[9]

In addition, of course, adding on a technology will involve a cost, over and above the cost of the manufacturing process. These add-on costs may well be lower than the costs of cleaning up environmental damage after it has been created. And in a well-regulated economy, such costs might also be offset against the costs of environmental fees for disposal or penalties for emissions. On the other hand, there may be circumstances in which the additional cost will threaten the productivity of the company operating the process.

Another worrying concern with end-of-pipe strategies is their overall effectiveness. Take the example of an end-of-pipe filter plant used to remove metal contamination emitted from a factory pipe into a river. By removing the heavy metal from the effluent, the filter plant attempts to protect the quality of the receiving water. We know from the conservation laws discussed in Chapter 1, however, that the metal itself persists in some form throughout the process. Even though it has been removed from the effluent, the same amount of metal remains in the sludge or filter-cake from the end-of-pipe plant. But what happens to this sludge? Sometimes, as we shall see later, it is possible to recover metal from the sludge. Sometimes, however, recovery is infeasible for physical or economic reasons. In this case we must look to different disposal options for the sludge. Treatment plant residues usually go either to landfill sites or to incineration plants. Neither of these options allows us to escape the conservation law. The metal must go somewhere. If it goes into a landfill site, there is a danger that the metal may eventually leak out of the site and into the water supply. If the sludge is incinerated, the metal might end up in the atmosphere and later be deposited on the ground or (once again) end up in the water.

One of the most perplexing features of this 'secondary' pollution is how difficult it is to control. Once a material has left the economic system it becomes subject to largely unpredictable environmental forces. The amount of metal leaking out of landfills or being emitted from incinerator stacks might be small. But the damage it causes depends on where it ends up. The uncontrolled contamination of precious groundwater supplies from leaking landfill sites may turn out to be an even more intractable problem than the controlled contamination of river estuaries.

Finally, we should recognise that the end-of-pipe strategy is effective only in reducing emissions from relatively large point sources. Increasingly, environmental problems are arising from a variety of **non-point** sources. These sources include run-off from agricultural land or from city streets, and the emissions associated with large numbers of individual consumers. The dissipative nature of the economic system has already been pointed out. This dissipative nature is reflected by the increasing number and geographical spread of the actors involved as we move along the industrial chain.

For any one nation, the primary processing of materials will occupy at most a few tens of large mining companies in specified locations. But secondary manufacturing facilities will number at least in thousands and be distributed throughout the country. Consumers – who constitute the last link in the materials flow – will number in millions. The end-of-pipe strategy is really only likely to be effective on specific emissions from large industrial sources. A number of other kinds of emissions – the dissipative uses described in the previous chapter, and the disposal of distributed material products – are much more difficult to address in this fashion.

PRUDENCE IN UNCERTAINTY

We can see from this discussion just how difficult environmental protection has turned out to be. One of the factors which contributes to this difficulty is the almost overwhelming complexity of the systems with which we are concerned: both the economic systems and the ecological systems. Another crucial factor is the enormous uncertainty which clouds our knowledge about these systems and their interaction with one another. Ironically, this uncertainty has often provided an excuse for disposal practices to continue unchecked for years, even when there was an obvious threat to environmental quality.

An example of this is provided by the disposal of wastes from the titanium dioxide industry. The industry produces non-toxic white pigments which were widely adopted as an environmentally acceptable alternative to the toxic lead- and zinc-based pigments which had been used previously. Despite the relative safety of the product, however, the manufacturing process for titanium dioxide generates hazardous liquid waste streams with high acid and metal content. For a long time the main disposal route for these wastes was dumping at sea.

From quite early on there was evidence of changes in the communities of marine organisms near the dump sites of the North Sea. Studies showed, for example, that there was an increase in epidermal papilloma (a form of skin cancer) among the local fish population. But for ten years no action was taken to halt the dumping practices because there was still uncertainty in establishing a causal link between the dumping of the wastes and the damage to the fish. The early studies focused on the acidity of the wastes as a potentially causative factor. Later it emerged that the likely cause was in fact the metal content of the wastes, with attention focused particularly on chromium. But the upshot of the continuing dispute was that known toxic wastes continued to be dumped, despite opposition from environmental lobby groups, for over a decade. Even when new evidence emerged of a correlation between chromium content in the fish and epidermal papilloma, scientists could still not prove an *irrefutable* causal link between the waste disposal and the fish disease, because the factors which govern the appearance of disease in any organism are extremely complex.[10]

This deadlock was only really broken by the emergence of what has since become known as the **precautionary principle**. Essentially this is a 'better-safe-than-sorry' principle which attempts to shift the burden of proof in disputes about environmental damage. Historically, the tendency has been to assume that industrial emissions are 'innocent until proven guilty' of environmental damage. The precautionary principle suggests that certain kinds of emissions should be regarded as 'guilty by virtue of their nature'. And attempts should be made to reduce such emissions, even in the absence of proof that they have caused particular environmental effects. After all, it is too late to worry about proof of a causal link *after* irreversible environmental damage has taken place.

THE EMERGENCE OF THE PRECAUTIONARY PRINCIPLE

The first international formulation of the precautionary principle was at the First International Conference on the Protection of the North Sea in 1984. In this context, the conference focused particularly on emissions into the marine environment. And it restricted its attention to substances which are 'persistent, toxic, and liable to bioaccumulate'. The conference called for emissions of these kinds of substances to be reduced 'even where there is no scientific evidence to prove a causal link between emissions and effects'.

Although the original formulation of the precautionary principle was restricted to the marine environment, we can see that exactly the same problems of complexity and uncertainty, and the same difficulty in providing irrefutable causal links can apply to all environmental media. So it would certainly make sense to apply the precautionary principle to emissions of all potentially hazardous materials.

What do we mean by potentially hazardous materials? The first formulations of the precautionary principle identified substances which were persistent, toxic and liable to accumulate in biological organisms and food chains. The events of Minamata Bay illustrate that these three properties particularly increase the risk associated with disposal practices.

But many other kinds of materials can be hazardous in the environment. Some non-toxic materials can be converted to toxic materials after they have been released because of the action of sunlight or bacteria; or because they interact with other materials in the environment. Synthetic chemicals have proved extremely damaging even when the substance in question is not hazardous to humans. This is illustrated clearly by the case of CFCs. These chemicals were introduced in the 1940s and 1950s to replace other chemicals such as ammonia which were known to be hazardous to human health. CFCs are themselves non-toxic to humans. But because they are persistent they have reached the upper atmosphere in large quantities. As we now know, they contribute to the depletion of the ozone layer.

This leads us to suppose that certain non-toxic chemicals should also be the object of careful scrutiny, and perhaps be subject to precaution. In fact, even those materials for which there are well-known natural cycles can cause major environmental problems. To see this, we need only consider the case of carbon dioxide emissions from human activities. No more 'natural' chemical compound could be encountered than carbon dioxide. It is a normal product of animal respiration. It is essential as an input to the process of photosynthesis. And the cycling of carbon through respiration and photosynthesis is essential to maintaining the balance of our atmosphere and climate. Yet carbon dioxide is also released when fossil fuels are burned; and the prospect of global warming from increased atmospheric carbon dioxide levels presents one of the most serious environmental problems with which we are faced today.

In fact, we should really say that there is a very wide spectrum of materials emitted from the industrial economy which pose potentially catastrophic and irreversible environmental threats. Some of these are highly toxic materials which do not even exist in nature. Others are naturally occurring materials for which well-known material cycles exist.

IS THERE A 'NO-WASTE' SOLUTION?

These examples seem to be leading us towards the possibility that *all* material emissions are potentially hazardous in the environment. In this case, should we not apply the precautionary principle to *all* material emissions? Should we be aiming for industrial systems which are completely 'closed' in material terms? In other words, should we be aiming for a human economy in which no material wastes ever leave the system?

The danger now seems to be that we are moving towards an unrealistic vision of what environmental management is to mean. Such a requirement is much stronger than anything imposed by nature. No natural ecosystem has 100 per cent internal material closure except the planet as a whole. In fact, younger ecosystems tend to recycle fewer materials than more mature ecosystems. More mature ecosystems may recycle more than half of the materials they use.[11] But this recycling is always less than 100 per cent. For the rest, the ecosystem relies on the global material cycles which are powered by the sun. This complex system of materials reuse is carefully regulated by the availability of solar energy. But nowhere in this careful scheme of things is there any suggestion of complete internal materials cycling.

The suggestion of complete materials closure of the industrial system also seems hopelessly demanding from an economic point of view. What is less obvious is that it is actually impossible from a thermodynamic point of view. As we have seen, thermodynamics characterises any material transformation as being dissipative of both energy and materials. The only way we could reverse the dissipation of materials from human activities would be by using vast amounts of energy which in their turn could only be obtained from material resources. We could come close to closing a particular material cycle if we supplied sufficient high-quality energy to the task. But supplying this energy is itself a thermodynamic process, unavoidably dissipating more materials and more energy. So thermodynamics makes it quite plain that we cannot simply stop dissipating materials into the environment from human activities. And it is clear that we would gain absolutely nothing from the suggestion that the economy should strive for zero emissions of all materials. The concept of a 'no-waste economy' is just an illusion. No such thing can exist.

On the other hand, no natural ecosystem actually achieves zero emissions either, and it is clear that many such systems have survived for very long periods of history. So perhaps we can learn something about environmental management from nature herself. Perhaps we can devise a new concept of environmental management which attempts to make the industrial economy more like a natural ecosystem.

TOWARDS PREVENTION

Concepts of environmental management have shifted, perhaps only during the last decade or so, towards the idea of acting **preventively** to reduce environmental emissions from human activities. Various names have been used to describe the emerging conceptual approach. These names include **waste reduction**, **waste minimisation**, **pollution prevention** and **source reduction**.[12] Later, phrases such as **clean technology**[13] and **clean (or cleaner) production**[14] were introduced to try and capture the idea of technologies and production processes which were inherently cleaner, emitted less waste and were less environmentally damaging than their predecessors. The principal characteristic of all these approaches was the emphasis on taking measures to reduce environmental emissions, and consequently reduce the need for expensive clean-up technologies, disposal sites, environmental remediation, and blind faith in nature's assimilative capacity.

The emerging environmental strategy has two main avenues to it. The first is to reduce the material intensity of economic activities. That is: for each activity carried out, fewer materials are used, fewer materials flow through the industrial process, and consequently fewer materials end up in the environment. If this reduction is to occur without losing the value of the service which those material flows provide, then what we are talking about is **improving the material efficiency** of providing different services.

Given the broad range of materials over which environmental concern exists, these efficiency improvements need to be applied quite generally to all the material flows in the industrial economy. On the other hand, there are certain kinds of materials which represent particular priority hazards in the environment. And this fact suggests that we might sometimes wish to substitute for one kind of material – a priority hazard material – another material, considered less hazardous. **Substitution** is therefore the second main avenue for a preventive approach to environmental management. It is important, however, that we should recognise that substitution does not only mean substituting one particular input material for another. Sometimes we may want to substitute one product for another. At other times we may need to substitute one whole industrial process for another. And occasionally, it may be legitimate to ask whether we can substitute one whole set of industrial activities for another.

We can see that, in a way, both these avenues are aiming at the same thing: to make the economic system look more like a natural ecosystem. The pathway of improving material efficiency appears to be one of the evolutionary progressions within natural ecosystems. At any rate, improvements in material efficiency can reduce the burden placed by the industrial ecosystem on the global ecosystem. Fewer materials flow out of the system. The burden of human intervention in the natural material cycles is reduced.

Substitution also has a role to play in developing a new **industrial ecology**. Some hazardous materials do belong to natural material cycles. Toxicity is certainly not restricted to the industrial economy. The toxic heavy metals, for example (such as mercury, lead and cadmium), have quite specific cycling patterns associated with the weathering of rocks, evaporative processes and patterns of sedimentation in rivers and oceans. But these cycles are relatively limited, and now subject to distortion by human activities.

In general terms, therefore, we should regard substitution away from particular kinds of hazardous materials as the prudent course of action. In particular, substitution away from dissipative uses of synthetic chemicals or materials which are toxic, persistent and liable to accumulate is a key strategy for a new approach to environmental management.

SUMMARY

To summarise the lessons of this chapter, we could say that environmental management has appeared in the industrial economy at best as a sort of afterthought, an addendum to the main business of producing economic goods and services. But the dangers evident and the failures witnessed by proceeding in such an *ad hoc* fashion lead us to a radically different assessment of the situation.

The message from past environmental failures and from the precautionary principle itself is that future environmental management needs to be built on **environmental foresight**. We can no longer hope that *laissez-faire* emission of materials from the industrial economy will be an acceptable strategy. Nor can we rely on the ability of the environment to assimilate all kinds of material emissions from the economy. Rather we must make continued and strenuous efforts – within the constraints of thermodynamics – to reduce the material impact of the economy on the environment.

What emerges from this preliminary analysis is that environmental management is a process which demands the utmost care and consideration. It is amongst the most difficult of the tasks which face the industrial society, if we are to navigate safely between the conflicting demands of our own needs and the constraints of the physical world. First and foremost, this task demands that we design and, where necessary, redesign environmental foresight into industrial processes, consumer products and material consumption patterns. At the same time, it is essential that we do better than we have done in the past, if industrial society is to bid farewell to the environmental negligence epitomised by the incident at Love Canal.

4

A STITCH IN TIME
The principles of prevention

INTRODUCTION

The idea of a preventive approach to problem-solving can be illustrated by reference to preventive health care. Curative medicine attempts to correct imbalances and diseases in the organism through surgery or through drugs. Preventive medicine seeks to prevent illness itself by promoting health in the patient, and increasing his or her natural resistance to disease. But prevention has to act upstream, as it were, in advance of the onset of disease. Once the illness has set in, the organism is already out of balance. Curative medicine can of course still 'prevent' a sick patient from dying, and often aids recovery. But it is generally more expensive and often more difficult than ensuring that the patient stays healthy to start with.

Preventive environmental management also requires that we act upstream, before there are environmental impacts. In fact, the further upstream we act, the more likely we are to be successful in preventing environmental pollution, as we shall see. Clean-up strategies can sometimes 'prevent' environmental emissions from impacting on human health. But they do not address the cause of those emissions. End-of-pipe strategies can 'prevent' the emission of specific pollutants into a particular environmental medium. But they do not address the generation of polluting emissions at the source.

In a sense, we could say that **prevention is a directional strategy**. It looks as far as possible upstream in a network of causes and effects. It tries to identify those elements within a causal network which lead to a particular problem and then takes action *at the source* to avoid the problem. First of all therefore, preventive environmental management requires us to identify the root causes of environmental emissions within the industrial economy. Next, it requires us to address those causative elements in ways which will reduce the environmental outputs. This chapter is devoted to a discussion of what these two critical tasks mean in practice.

THE ECONOMY AS A PROVIDER OF SERVICES

Chapter 1 presented a picture of the industrial economy in material terms. Figures 1 and 4 illustrated the essentially linear flow (from resources to waste) which constitutes the material basis of the economy. This linear flow is largely responsible for the environmental concerns which we now face. At the same time I pointed out that the aim of these material flows was to supply the demand for goods and services from **consumers**: households and individuals. Although there is a traditional distinction between goods and services, a little reflection reveals that material goods are themselves also required *mainly* to provide certain services.[1]

Let us take as a starting point therefore the idea that **the economy is a provider of services**. Figure 9 presents a simple summary of the material dimensions of this idea.

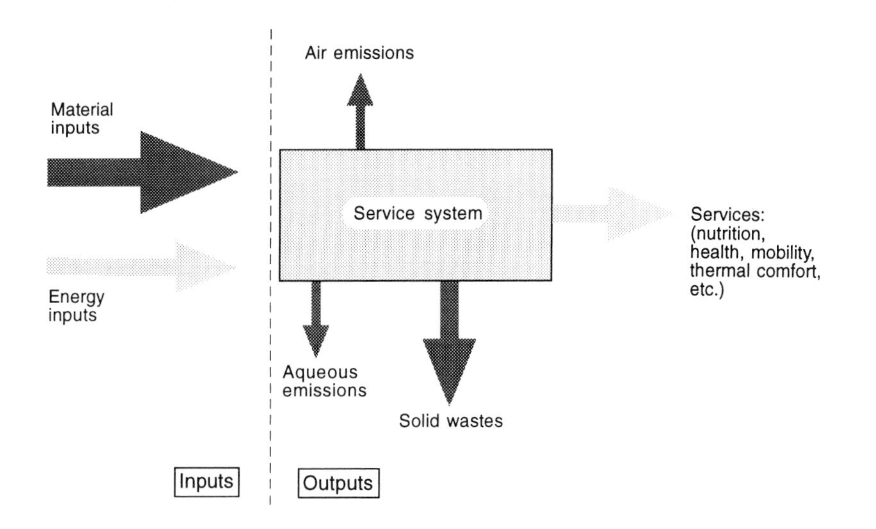

Figure 9 Material flows from the provision of services

The system itself is presented in Figure 9 as a kind of 'black box'. There are certain material inputs and certain material outputs. But the main *functional* output from the system is the service which it provides. This could be a single service such as nutrition or health or mobility and so on. Or we could envisage the black box as a combined system with a variety of service outputs.

From a thermodynamic viewpoint, the concept of providing services implies performing some kind of useful work. We know from Chapter 1 that this useful work can only be carried out by accessing available energy in some form, and employing high-quality material resources. So we know that the black box always requires energy and material inputs. The system then processes these inputs in such a way as to provide the required service or services. Since there are material inputs to the system, there must also be material outputs. Again using the discussion of Chapter 1, we know that these material outputs are in a degraded form: energy which is less available to us, and materials which tend to be more dissipated and less useful than the input resources. In this diagram they are portrayed as waste emissions of various kinds.

THE PREVENTIVE STRATEGY

The preventive strategy seeks to identify the causes of environmental emissions and find ways of reducing these emissions. At one level, we could take the **demand for services** as the underlying causative element. But as I have already remarked in Chapter 3, we must draw the line somewhere between reducing material throughputs and jeopardising human welfare. Eliminating the demand for services is not a legitimate or viable basis for any environmental management strategy! This does not mean that the demand for services should remain entirely unchallenged. But it is a complex issue, and requires a careful and balanced understanding of where those demands come from. I shall return to this topic at later stages in the book.

For the moment, let us concentrate our attentions on the black box itself. One way of characterising earlier approaches to waste management is to say that they focus all their attention on the material emissions from the given system. They take for granted both the demand for services and the complex technological parameters of the black box itself.

The first point to make about the preventive approach is that it must seek to unravel the hidden complexity of the service system. It is *within* this system that material input requirements are decided on, and material outputs are determined. To apply the preventive strategy effectively, therefore, we must first look within the system and then be prepared to redesign – and where necessary reconceive – it altogether. So let us now take a closer look at the component parts of the system illustrated in Figure 9.

Figure 10 breaks the service system down into several stages which are more or less sequential. The rectangular boxes in the diagram represent the familiar stages of material transformation within the economy: mining, manufacturing and distribution of material products, and subsequent waste management after use. These stages require resource inputs and they deliver material outputs. The oval stages are essentially non-material in nature, although they are intimately related to the material aspects of the system.

The first and perhaps the most critical of these stages is the one which represents the **conceptualisation** of the provision of a particular service. It is here that the performance parameters of the system and its material dimensions are ultimately determined. For instance, we can conceive of a system for providing food which is based on a global marketplace, chemical fertilisers, advanced agricultural machinery, and a complex transportation network. Or we can conceive of a system which is based on locally produced, labour-intensive, organic[2] farming methods. Each of these systems will have very different material implications.

The next stage in the representation is the **design** phase. This is the stage during which all kinds of aspects of the conceived system are planned and designed. In particular, of course, this must include the design of both products and production systems. Both of these elements will have significant impacts on the material aspects of the overall system.

Subsequent stages in the diagram represent the **material processing** and **production** stages discussed in Chapter 1. These stages produce the **material products** which are then distributed to the consumer to provide specific services. Having served their purpose, the degraded products then pass to a further material stage, which I have called here **waste management**. After passing through the waste

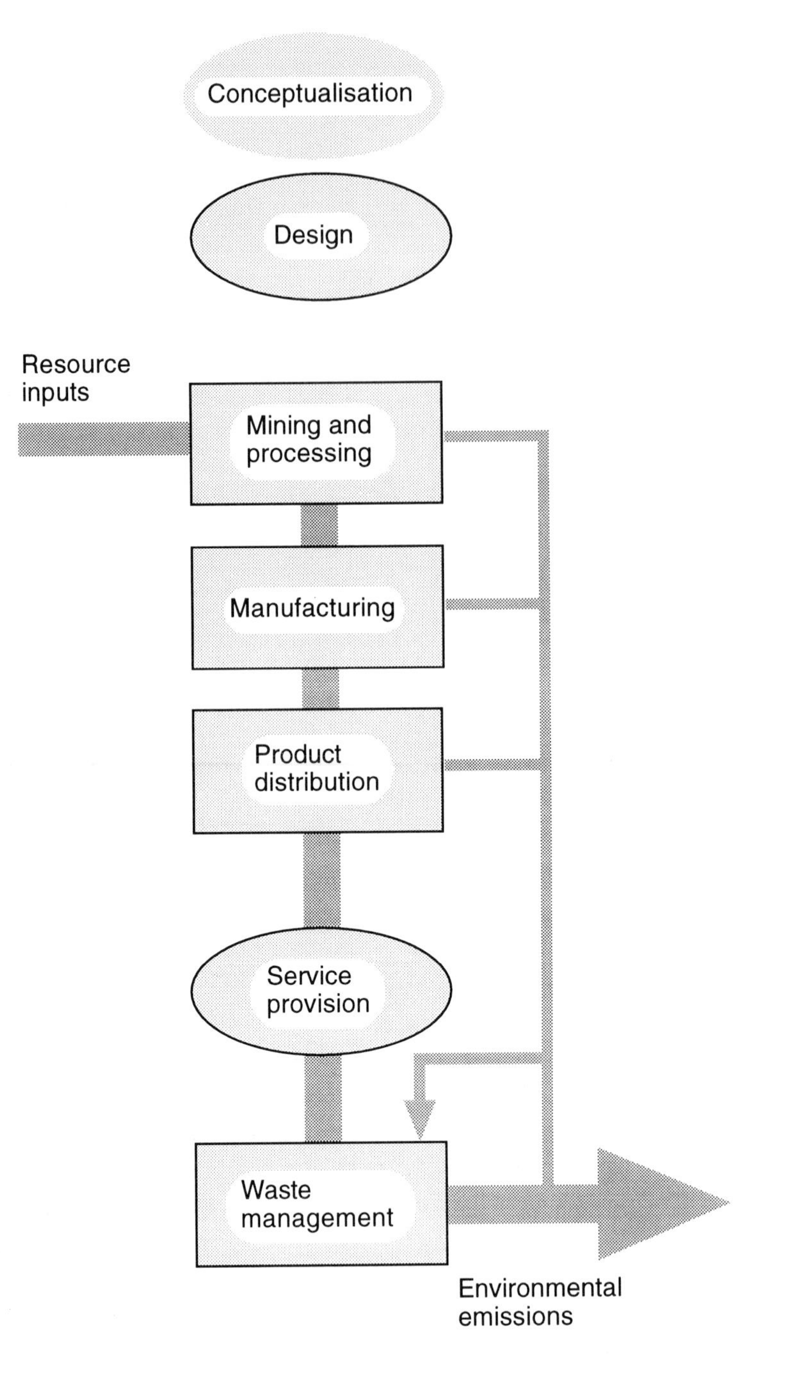

Figure 10 Structure of the service system

management sector, materials will leave the economic system and re-enter the environment.

Often this last stage in the life of products is referred to as **disposal**. But this stage is not just a question of throwing materials away and forgetting about them. The product will depart from the economic system and even from our consciousness. But we should always remember that materials that we have disposed of continue to be subject to the laws of nature. Sometimes they degrade; sometimes they disperse; sometimes they accumulate; sometimes they rejoin material cycles of one kind or another; and sometimes they leak into soils and water supplies, causing environmental damage for generations to come. This is one of the prime motivations for developing a preventive environmental management strategy.

The preventive strategy itself is composed – as I described at the end of the previous chapter – of two principal avenues of intervention in the service system:

1 **improvements in the material efficiency** of the system; and
2 **substitution** for hazardous throughputs.

We could also identify another important general strategy, which is to **reduce mixing** of different kinds of materials within the system. From a thermodynamic viewpoint this mixing represents an increase in entropy within the system. From a practical point of view, mixed waste streams are much more difficult to deal with effectively than segregated streams.

THE TIMING OF PREVENTION

Figure 10 presents the different stages in the provision of services as sequential. In reality, of course, there is no strict temporal order. It was already noted in Chapter 1, for instance, that some of the material outputs of the secondary (manufacturing) sector are destined for use in the primary sector or in the secondary sector itself. Also, there is inevitably a sort of feedback process in operation between the service provision stage and the design stage. The emergence of new demands and the decline of saturated markets will lead automatically to product innovation and design changes.

Perhaps most importantly, in a modern market economy **there is no single identifiable process of conceptualisation** within a definite temporal framework. Rather, conceptualisation is something that occurs continually throughout the complex operation of the system as a whole. This is not to suggest that conceptualisation cannot be influenced within the industrial economy. In fact, as we shall see below, it is by intervening during the conceptualisation process that the preventive strategy can have its biggest impacts. But conceptualisation is a rather broad interactive process involving many different parties and occurring over a longer timescale and in a more complex fashion than is suggested by Figure 10.

At the same time, the simplicity of Figure 10 highlights an important distinction between previous environmental management strategies such as dilute-and-disperse or end-of-pipe management and the emerging preventive approach. This difference is one of timing. Or – if the strict temporal nature of the system is denied – we could say that the difference is one of 'placing'. Figure 11 illustrates this difference.

The earliest *laissez-faire* attitude to the environment really involves no action at all during the provision of services. But it often leads to the need for remedial measures to be taken after environmental damage has occurred. The dilute-and-disperse and the containment philosophies both involve taking action at the stage of waste management. But these actions are often limited to controlling the rate at which pollutants enter the environment or to fortifying landfill sites. The end-of-pipe approach requires intervention mainly at the manufacturing stage.

By contrast, the preventive approach is about foresight. It must therefore operate both during the conceptualisation and during the design of service systems. The importance of appropriate design can scarcely be overemphasised. As the US National Academy of Engineering has pointed out: 'Design should not merely meet environmental regulations; environmental elegance should be a part of the culture of engineering education.' The different elements of designing preventive environmental management into the provision of services are discussed in the following sections.

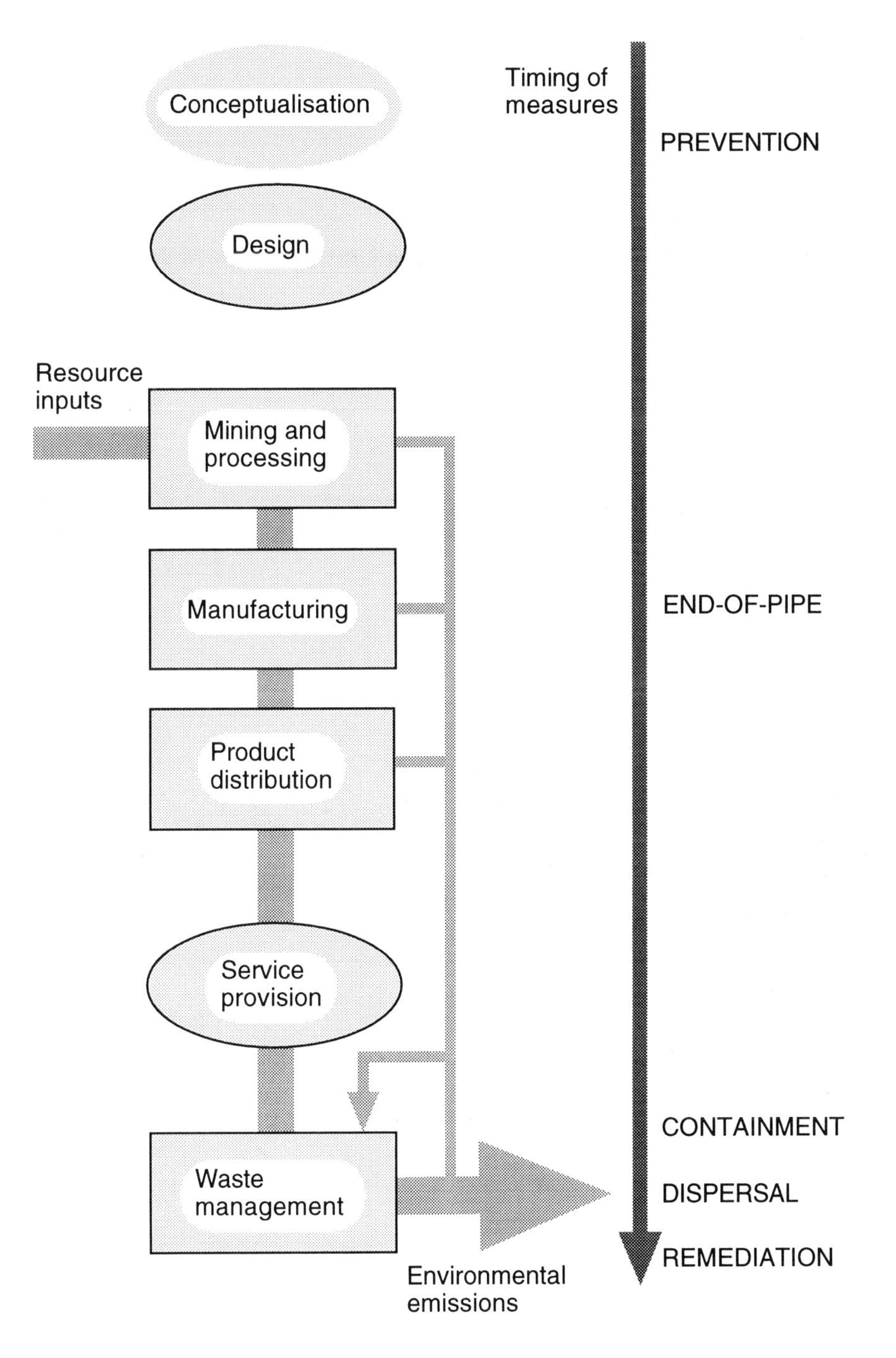

Figure 11 Timing of environmental management strategies

DESIGNING INDUSTRIAL POLLUTION PREVENTION

Much of the early initiative in pollution prevention focused on redesigning industrial processes to reduce the generation of process wastes. It is clear from the previous discussion that this cannot be the limit of our attempts to design environmental elegance into the system of provision of services. But it is certainly an important part of it. There is now a considerable body of experience and expertise in this important area. So it is worth while summarising briefly the key elements in industrial pollution prevention. Let us start out by looking at a few simple flow diagrams.

Figure 12 provides a simple schematic diagram of an industrial process. Like the service system depicted in Figure 9, the industrial process is represented in Figure 12 as a kind of black box. The main difference between Figure 9 and Figure 12 is that the functional output in Figure 12 is a material product rather than a service. These material products are manufactured from raw material inputs to the process. Only a proportion of the total material input is turned into useful product. The balance appears in the form of material wastes. Some materials are emitted to the air; others to the water; and some materials end up in the solid waste stream.

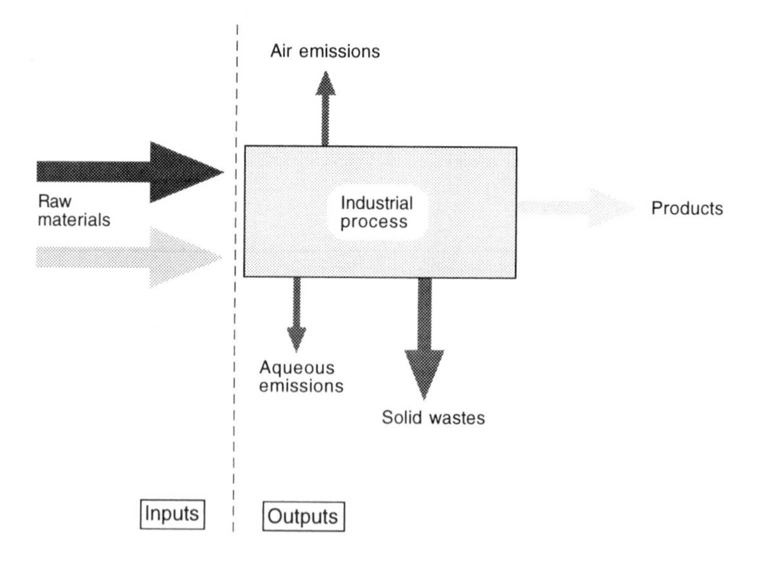

Figure 12 Simple industrial material flows

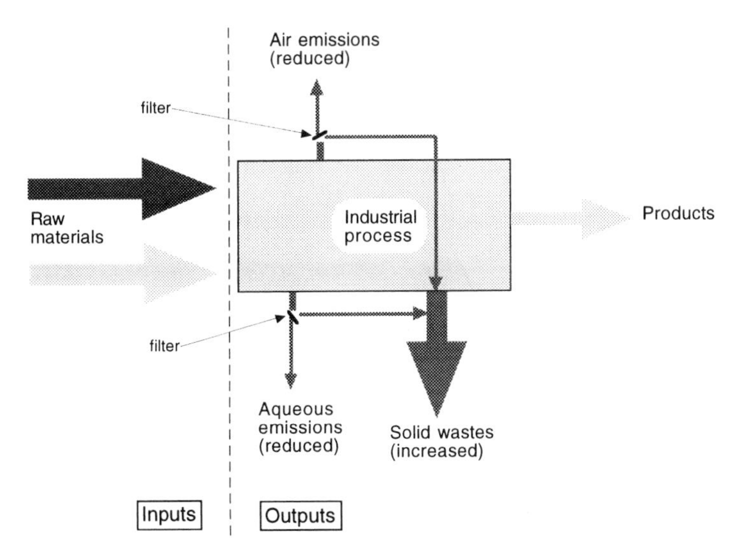

Figure 13 End-of-pipe treatment of air and water emissions

The mass balance principle discussed in Chapter 1 means that the total emissions to air, water and solid wastes are equal to the difference between the raw material inputs and the product outputs. And this mass balance principle applies both at the level of total mass, and also at the level of individual molecular contaminants.

End-of-pipe strategies (Figure 13) are usually aimed at reducing specific air emissions and waterborne effluents. They take out particular contaminants by adding filters, scrubbers and treatment plants on to the industrial process. Although they may be successful at reducing air and water emissions, what typically happens is that solid waste emissions increase. The simple mass balance equation makes it clear why this is the case. If the total raw material inputs are the same and the product output is constant, the balance between them must also remain unchanged. This unchanged balance constitutes the total waste emissions. If the air and water part of these total material emissions is reduced, then the solid waste emissions must increase to keep the total constant.

There are a number of variants on the end-of-pipe approach shown in Figure 13. For instance, some air-emission control technologies employ flue gas scrubbing, in which soluble flue gases are dissolved in jets of water. This process generates an increase in the aqueous

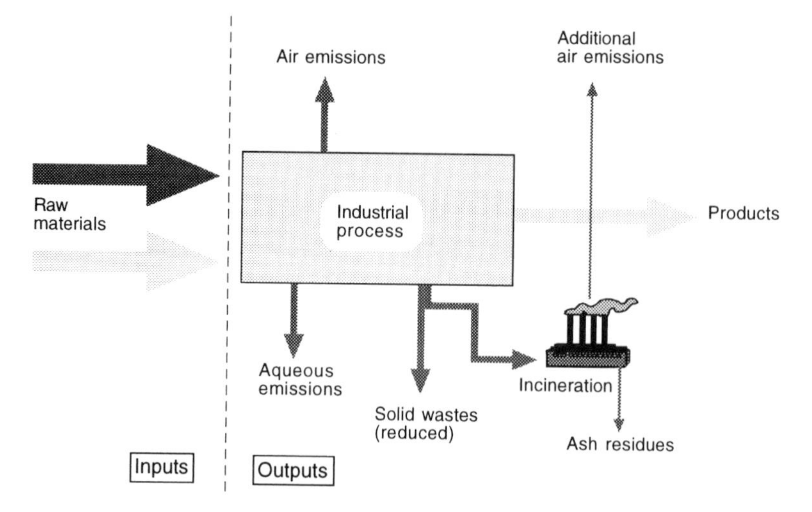

Figure 14 End-of-pipe treatment of solid wastes

emissions rather than in solid wastes, unless the water is subsequently treated and recycled. Another variant (Figure 14) exposes solid wastes to some kind of end-of-pipe treatment, such as incineration. Incineration reduces the total mass of solid wastes. But it does so at the expense of generating new air emissions.

The contrast between the prevention strategy and the end-of-pipe abatement strategy is illustrated by Figure 15. The mass balance principle is still observed. The product output has not changed. But now reduced material emissions correspond to reduced material inputs. If we were to represent the material efficiency of the process by the ratio of the product output to the material outputs, we could say that the material efficiency of the process in Figure 15 is improved over that of the processes in Figures 12 to 14.

Figures 12 to 15 still represent the industrial process itself as a more or less opaque black box. In reality, of course, it can be an immensely complicated network of material flows and technological conversion processes. And it is really only by delving into this network – which differs from process to process and from industrial sector to industrial sector – that we can realise the opportunities for industrial pollution prevention. It is only by making specific changes within the industrial process itself that we can improve material efficiencies and reduce the

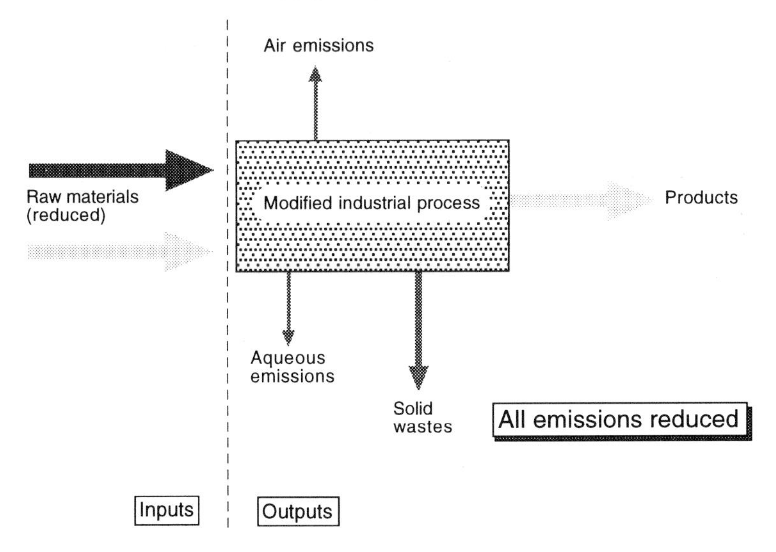

Figure 15 Industrial pollution prevention – improved material efficiency

burden of material emissions. It is beyond the scope of this book, and probably beyond the power of any book, to present a detailed encyclopaedia of technological opportunities for pollution prevention in every industrial sector. Nevertheless, it is certainly worth outlining a few key strategies which lead to this kind of change, and illustrating these strategies with examples from real life.

SIX KEY STRATEGIES FOR INDUSTRIAL POLLUTION PREVENTION

The first key strategy of industrial pollution prevention is the implementation of **good housekeeping** measures. This means generally improving the way in which materials, particularly hazardous ones, are purchased, stored, conveyed, handled and used in the industrial facility. Good housekeeping means identifying and reducing leakage and spillage, carrying out regular maintenance of all materials processing equipment, and instituting better inventory controls. A particular prerequisite for good housekeeping is to implement regular **waste reduction audits** (Box 1).[3] These are systematic accounting processes which track materials flows through the process, monitor

BOX 1 ELEMENTS OF A WASTE REDUCTION AUDIT

A waste reduction audit is a systematic, periodic, internal review of a company's processes and operations with the aim of identifying and providing information about opportunities to reduce wastes. The following six stages can be identified.

1 AUDIT PREPARATION

- prepare and organise the audit team
- advise personnel
- identify management responsibilities

2 MATERIAL BALANCE

- identify processes and construct flow diagrams
- identify process inputs (energy, water, materials)
- identify outputs (products, by-products, waste emissions)
- identify recycle and reuse rates
- derive a preliminary materials balance
- evaluate and refine the materials balance

3 ECONOMIC BALANCE

- identify financial parameters: inputs costs, waste disposal costs, emissions charges, product revenues and so on

4 IDENTIFY WASTE REDUCTION OPPORTUNITIES

- identify pollution prevention measures: good housekeeping, input substitution, recycling, process change, clean technologies, product reformulation

5 EVALUATE OPPORTUNITIES

- undertake technical, environmental and economic evaluation of waste reduction options
- list and prioritise viable options

6 DESIGN A WASTE REDUCTION STRATEGY

- formulate strategies and measures
- produce timescales for implementation
- identify waste reduction targets

the efficiency and proper functioning of the process, and identify opportunities for waste reduction.

Another basic strategy is to implement **internal recycling**. In many industrial processes there are opportunities to collect materials after they have been used and recycle them for the same or another use within the process. This will reduce material emissions from the process. It will also reduce the need for a continued supply of raw material inputs. Examples of this kind of measure include the recovery and reuse of solvents and acid cleaning baths in electroplating, the recovery and refurbishment of catalysts, and the precipitation and recovery of metals from treatment sludges.

Both internal recycling and good housekeeping measures are examples of simple process modifications. **Other process modifications** may also contribute to industrial pollution prevention. Some of these other measures are simple. Some are more complex. Examples include: the use of electronic process controls to moderate and optimise material flows; the addition of minor capital investments (such as membranes and filters); and the segregation of waste streams to aid recovery and recycling.

A continuation of the idea of process modification is the implementation of newer, more efficient and **cleaner process technologies**. These are generally processes which differ from the original in that they possess inherently better material efficiencies or inherently reduced reliance on hazardous materials. Examples of such clean process technologies include: the development of membrane technology to replace electrolysis in the chlorine production industry;[4] and the employment of electrolytic (rather than thermal) smelting technologies.

Most of the techniques for industrial pollution prevention which I have described so far relate to improved material efficiency. This was the first of the two principal avenues for preventive environmental management which I have identified. The other avenue was substitution. And **input substitution** is another important technique for industrial pollution prevention.

Figure 16 illustrates the main idea behind input substitution in industrial processes. Hazardous raw material inputs are replaced with less hazardous materials. This substitution can have a number of different impacts. In the first place, particular hazardous components in the waste streams can be eliminated.[5] Next, the products themselves can be less hazardous, because they no longer contain that particular material. Last but not least, there are advantages to be gained in terms of safety in materials handling.

Figure 16 also illustrates the sixth strategy for industrial pollution prevention, that of **product reformulation**. In the particular example

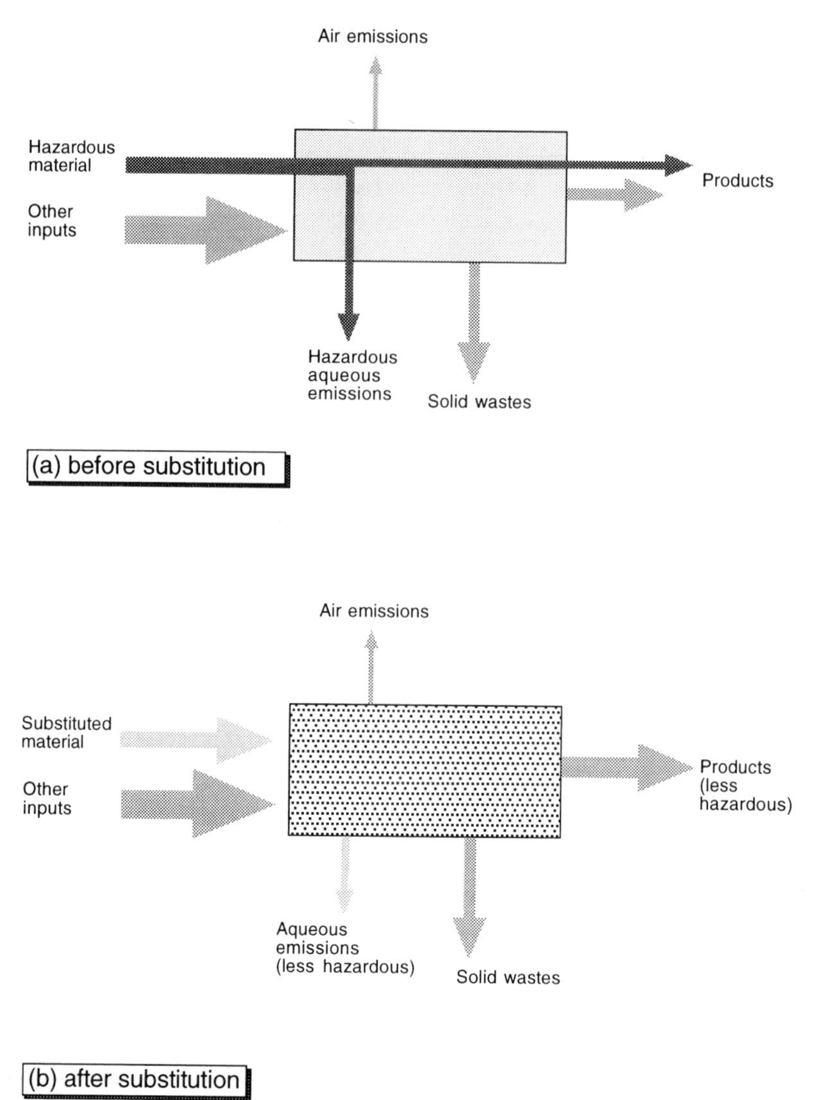

Figure 16 Industrial pollution prevention – input substitution

shown here, the product is reformulated by substituting less hazardous materials for its hazardous components. This reformulation also reduces the hazardous content of the aqueous waste stream. An example of this kind of reformulation is provided by the development of non-CFC-based aerosol sprays. This change means not only that CFCs are eliminated from the manufacturing process, but also that the product itself is less liable to harm the environment. Quite generally, reformulating the product can lead to multiple advantages: reduced material input requirements, fewer hazardous throughputs and emissions from the industrial process, and safer products.

In fact, we can see that the idea of reformulating and redesigning products is a very powerful one which goes a long way beyond simply reducing production process wastes. And we could say that the strategy of **redesigning products represents a higher level of prevention** because it starts to look even further upstream in the industrial chain – the demand for particular products and services. Ironically, product reformulation has been one of the least favoured options within industrial pollution prevention. This is partly because manufacturers are naturally concerned about loss of product quality. Sometimes, though, product reformulation has had rather dubious results because it has meant the transfer of hazardous substances from the waste stream to the product. Although this has the effect of reducing hazardous wastes from the facility, it clearly increases the hazard associated with the product itself.

THE DILEMMA OF SUBSTITUTION

There is a more general potential difficulty associated with any strategy of substitution. How do we know that we have substituted something which is better (from an environmental point of view) for something which is worse? How do we ensure that we do not simply substitute one environmental hazard for another? We have already seen that the range of materials which cause environmental concern is very wide. We have also seen how easy it is for an abatement measure simply to transfer pollution from one place to another. But we should not assume that this problem simply disappears when we are talking about preventive measures.

For instance, suppose that a particular factory reduces its use of a particular toxic compound by an internal recycling measure. We know from thermodynamics that energy is required to close a material cycle. This means that the factory is likely to be using more energy as a result of the measure than it was before. Suppose that this energy is provided by electricity. The toxic emissions from the factory itself will

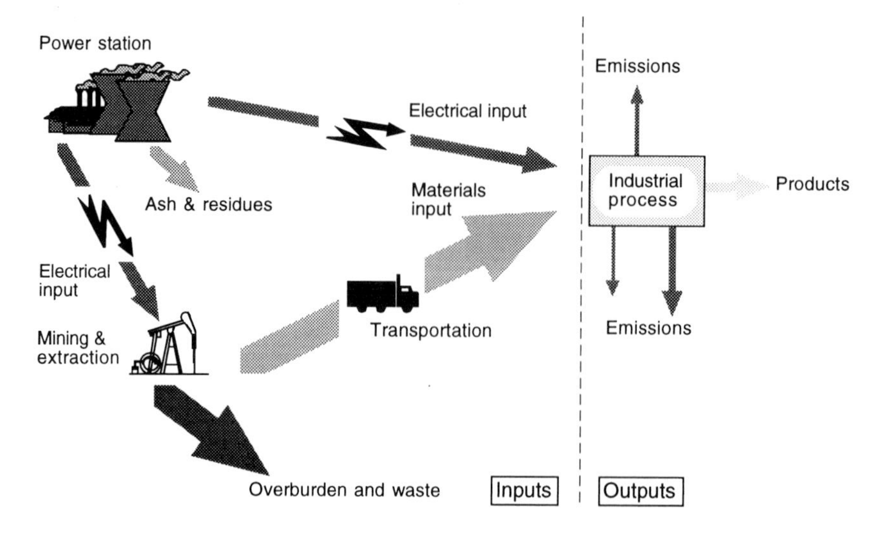

Figure 17 System effects of pollution prevention measures

certainly be reduced by the recycling measure, but more electricity is needed to close the cycle. So, in a sense, we could say that we have substituted electricity for the toxic compound in question within the industrial process. Since we know that the generation of electricity is a polluting activity, we could also say that we have substituted toxic pollution with sulphur pollution and carbon pollution (if it is a coal-fired electricity generation plant, for instance).

This is not the end of the story, however, because by recycling the toxic compound we have reduced the material input requirements. Since energy is required to provide these raw material inputs, the internal recycling measure can also be said to have reduced the energy requirements on the raw material side.[6] What I am drawing attention to here is that each industrial process is a part of a larger system. What we described simply as the 'input side' of the industrial process in Figure 12, turns out (Figure 17) to be a complex interrelated system of material interactions, each involving different raw material requirements and different kinds of environmental emissions. The question of whether there is an overall increase or an overall decrease in environmental emissions can only be resolved by looking at the wider system of which the industrial process is a part.

This system aspect of industrial production is just one of the factors which makes it difficult to decide whether a particular substitution

BOX 2 ELEMENTS OF LIFE-CYCLE ASSESSMENT

Life-Cycle Assessment (LCA) is a process of evaluating the environmental burdens associated with a product, process or activity by identifying and quantifying energy and materials used and wastes released to the environment; of assessing the impact of those energy and material uses and releases to the environment; and of identifying and evaluating opportunities of effecting environmental improvements.

LCA has the following five stages:

1 GOAL DEFINITION

- define the system under consideration
- identify the system boundary
- identify the purpose of the assessment (policy, engineering, economic)

2 INVENTORY

- identify and quantify resource requirements
- identify and quantify environmental emissions

3 CLASSIFICATION

- assess the contribution of resource requirements to resource depletion
- assess the contribution of environmental emissions to environmental burdens

4 VALUATION

- assess the relative importance of environmental burdens and resource demands
- assess the reliability of results and sensitivity to key parameters

5 IMPROVEMENT

- use the results of the exercise to identify environmental improvements, use of new technologies, or changes in practice

makes us better off or worse off in environmental terms. Another factor is the difficulty which we are likely to encounter in evaluating trade-offs between different kinds of environmental emissions. Is the environmental benefit associated with a decrease in toxic emissions greater or smaller than the environmental burden associated with increased carbon dioxide emissions from an electrical plant? How can we measure the relative environmental impacts of different kinds of emissions? A considerable amount of effort is now being dedicated to the development of **life-cycle assessment** methodologies (Box 2) which attempt to quantify and assess the environmental burdens associated with different technologies and processes. But even using such techniques, none of these questions is very easy to answer.

The whole issue becomes much easier when we can say unequivocally that a particular pollution prevention action has reduced *all* of the material inputs – as well as all the waste emissions – associated with the system in which the industrial process is embedded. In this case, of course, we are really talking about overall material efficiency improvements again. And the only caveat we need to be aware of is that there are thermodynamic limits to such improvements. These limitations arise because of the second law. Materials and energy are degraded during transformation. In order to return them as useful inputs to further transformations they must be upgraded and this can only be done by supplying more high-quality energy.

At the moment, in the existing industrial system, we are still a long way from thermodynamic limits. Improvements are possible in all kinds of industries and processes. Many of the most obvious efficiency improvements relate to the use of energy. The possibilities for saving energy in industry (for instance in process heating applications and in motive power) are now well-documented.[7] But I hope it is clear from the preceding discussion that we still need to make a careful assessment of the impacts of each process modification, each new design concept, and each substitution.

ADDRESSING PRODUCT POLLUTION

Let us now shift our attention from the industrial process to the product. Environmental burdens arise as much from what leaves the factory gate as from what emerges from the factory pipeline. And environmental management strategies have to take account of the product in two very specific ways. First, it is the demand for products which drives the industrial processes, and hence provokes the environmental emissions from industry. Second, and equally importantly, products also pollute. In particular, when they are thrown away they result in the ever-increasing quantities of post-consumer waste with which industrial societies are having to cope.

Clearly, therefore, environmental management should not confine itself to industrial processes. Even if much of the focus of the pollution prevention work of the 1980s has been on industrial processes, we must somehow address the product itself if we are to get at the roots of environmental degradation. But what can we do about product pollution? Products wear out. The laws of thermodynamics tell us that they must wear out. Materials become degraded and energy becomes dissipated as the result of activity in the economic system.

Although we cannot escape completely from this inevitable degradation and dissipation of materials, we can look for opportunities to reduce it to a minimum. In particular, we can do this by ensuring that we get the **maximum possible use** out of material products, by preventing needless dissipation into the environment, and by preserving and protecting products from unnecessary decay and obsolescence. This overall strategy has been called **product-life extension**.

EXTENDING THE LIFE OF PRODUCTS

The key aim of product-life extension activities is to extend the period of time over which products fulfil useful functions in the industrial economy. In brief, we aim to get as much use as possible out of materials before we throw them away into the environment. This will not only reduce the burden of post-consumer waste. It will also reduce the demand for new material products, and reduce the environmental burdens associated with producing them.

The key elements of extending the utilisation period (see Figure 18) are the following:[8]

1 **reuse** of products either for the same purpose or for another purpose;
2 **repairing and maintaining** products to keep them useful for as long as possible;
3 **reconditioning** or **remanufacturing** products to restore or upgrade them;
4 **recycling** of raw materials from products to provide material inputs to the manufacture of other goods.

It is important to remember that each of the four material loops illustrated in Figure 18 involves new processes of material transformation, each of them subject to the laws of thermodynamics. Consequently, each of them is likely to require energy – and perhaps also material – inputs. These new inputs will involve new waste emissions into the

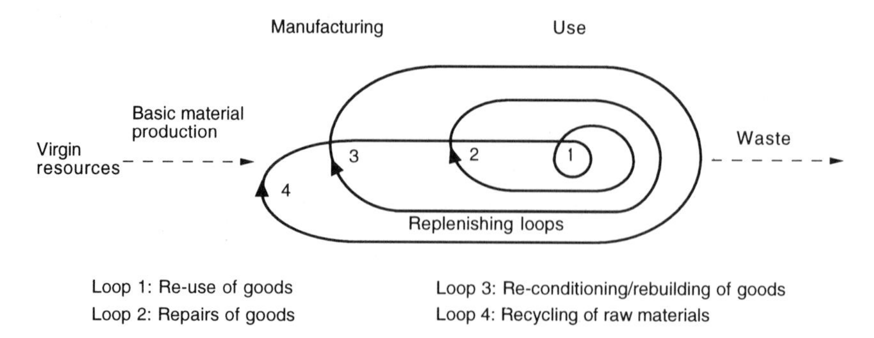

Figure 18 Product-life extension strategies

environment. In many cases, these new waste emissions will usually be very much less than the waste emissions which would have been created by making brand new products and just throwing the old ones away. Once again, however, we need to make a careful examination of the whole system before we can decide whether a particular strategy for reusing a material or a product is less or more environmentally damaging than making a new product or using new materials.

What is generally found is that there is an approximate 'hierarchy' amongst these different strategies, based on the relative amounts of energy and materials that are needed to carry them through. The strategy of reuse is at the top of this hierarchy because it generally requires least additional energy and material input. Usually this additional energy and material input is required for collection and redistribution of the reusable products. Where there is a well-established local network for collection and redistribution, the additional energy demands tend to be minimal.

Repair and maintenance activities usually require a greater input of energy and materials than reusing products directly. The success of the repair and maintenance loop really depends on an appropriate infrastructure to provide both parts and labour for maintenance work.

Reconditioning generally requires somewhat higher inputs of energy and materials, and may involve the setting up of centralised facilities for remanufacturing. Products are dismantled or stripped to separate worn out materials from those which are still useful. The useful parts are then reconstructed using new materials where necessary to make new products. This process offers the possibility of **technological upgrading** to improve product performance as well as lengthening product life.

During the recycling loop material products have to be collected and reprocessed to separate out raw material components. These raw materials can then be used as inputs to the manufacture of new products. The energy of collection, separation, treatment and redistribution can make the recycling loop the least efficient of the loops from a materials perspective. Nevertheless, the energy and materials required for recycling may often be less than those required to extract and process primary raw materials.

The overall effect of these strategies is to introduce a kind of **cascade of use** for material products.[9] New products made from virgin materials enter the system at the 'top' of the cascade. For a while these products may be reused directly with minimal additional inputs of material and energy. But at a certain point, they will require more substantial repair and maintenance. Following this repair process, products may re-enter the system at a lower point in the cascade – perhaps corresponding to a lower economic value.[10] Later the same products may require full-scale reconditioning which returns them to a new state of usefulness – perhaps even their original state. Eventually, however, material products move down the cascade. Towards the bottom of the cascade the only value in the products lies in their component materials. These materials must then be recovered and recycled as inputs to another cascade of use.[11]

By supplying services at each of a variety of different levels, this cascade maximises the **use value** derived from each material input. The quantity of virgin materials entering the system is consequently reduced. So are the environmental emissions. But the same level of service is maintained (Figure 19).

The picture painted in Figure 19 is clearly analogous to the one illustrated in Figure 15. Reduced material inputs are required to produce the same output. In a sense, therefore, this strategy once again corresponds to an improvement in material efficiency. This time we are talking about the efficiency with which materials are used to supply particular services rather than products, but the basic idea is the same. Instead of talking about material efficiency, we could equally talk about reducing the **material intensity of providing a particular service.**

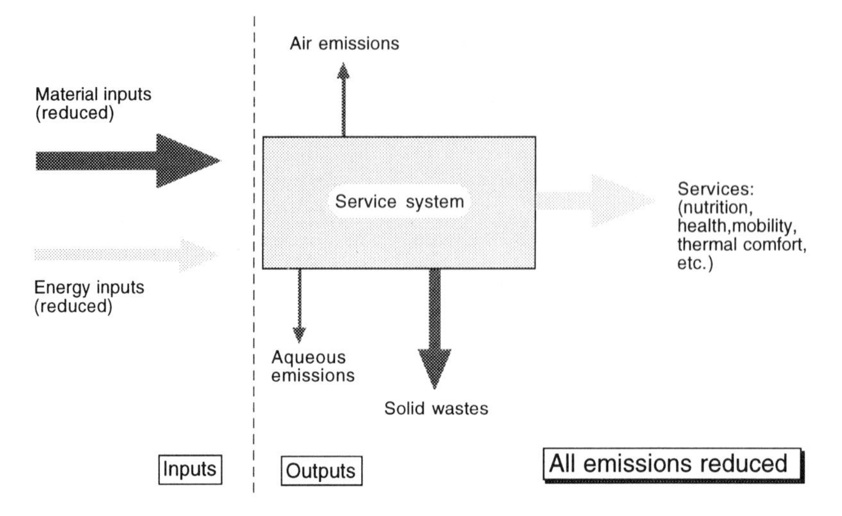

Figure 19 Providing services: improved material efficiency

Improved material efficiencies then correspond to reducing the material intensity per unit of service.[12]

Much of the success of these strategies depends on the existence of an appropriate infrastructure within the industrial economy. For example, we need decentralised collection and distribution networks to reduce the transportation needs of reuse and recycling. Under these conditions product-life extension becomes considerably more attractive, leading to reduced environmental burdens and lower economic costs.

PREVENTION AND THE PROVISION OF SERVICES

It should be clear from these discussions of industrial pollution prevention and product-life extension how important the design and conceptualisation stages are to a successful preventive strategy. In fact, design for long life could be considered a high-priority strategy for optimising utilisation of resources in its own right. Different aspects of design are important. Materials selection is clearly crucial: materials can be selected for improved durability; they can also be selected for recyclability. In addition, however, material selection will have important implications for component replacement. And modular design which allows for the replacement of component parts – rather than throwing away the whole product – is another important aspect in designing long-life products.

Ultimately, however, it is the way in which the provision of services is *conceived* which can have the biggest impact on the material intensity of the system. In Chapter 7 we shall have occasion to see just how extensive the opportunities for reducing material throughput are once we begin to reconceive the provision of particular services.

It is important to note here that **services are not material outputs** from the system. We cannot measure a service in terms of tonnes of a certain material or even the quantity of goods provided. Rather, services are composed of a variety of different factors such as thermal comfort, nutrition, mobility, health care, recreation and so on, each measurable only in units relevant to that service. The quantity of material goods consumed is often used as some kind of proxy for services provided. In some cases this correspondence may be valid. For instance, there is some correspondence between the provision of foodstuffs and nutrition. On the other hand, there are clearly cases where the proxy is not particularly useful.

Let us consider a particular example. There is now a well-developed system designed to supply consumers in the industrial economy with delivered fuels: coals, oils, gas, electricity. But these fuels are of no use to consumers on their own; they are useful only because they can provide certain **energy services**: thermal comfort, for example.

Generally speaking, however, we can provide the same level of energy service in a number of very different ways. For instance, we can provide thermal comfort in a draughty house by using an open and very inefficient coal fire, and burning a large quantity of coal in it. Equally, we could provide the same measure of thermal comfort by insulating and draught-proofing the house, and installing a high-efficiency boiler to burn oil or gas. A much lower quantity of material fuels would be consumed, but the same level of service (thermal comfort) would be provided.

In fact, we could reasonably take the example further than this. I have really taken the expression 'thermal comfort' to refer to a particular room temperature, and this is usually the way in which thermal comfort is measured. But if we are to do any justice to the complex metabolism of the human body, we should recognise that the relationship between thermal comfort and ambient temperature is not a simple, direct one. As the room temperature in a cold dwelling increases, the thermal comfort level also increases up to a certain point. After that point, however, rising room temperature rapidly indicates thermal discomfort!

What is perhaps even more interesting is that this changeover temperature, at which thermal comfort becomes thermal discomfort, is a variable point and not a fixed one. For instance, thermal comfort can be maintained in a much colder room by wearing an extra sweater. In a warmer room, the same attire would contribute to thermal discomfort. Sweaters are clearly material products, of course. And by ensuring thermal comfort by lowering room temperature and providing sweaters, we have substituted material sweater resources for material fuel resources. Only by looking at the larger system impacts could we decide whether we are environmentally better off wearing sweaters in colder rooms or raising the room temperature and going naked.

At this point, I do not want to delve into the complex issue of personal choice, which also clearly affects our decisions about thermal comfort. This is the subject matter for a later chapter in this book. But I hope that the point of the illustration is clear anyway. Thermal comfort and room temperature do not bear a simple proportional relationship to each other.

There is another interesting lesson from this case study. The human metabolism is such that thermal comfort depends intrinsically on our physical activity levels. When we are largely inactive – for instance when we are asleep or while we are engaged in predominantly sedentary occupations – thermal comfort requires a higher ambient temperature than when we are actively engaged in physical exertion.

This simple physiological fact raises the possibility of reducing our space-heating requirements by changing our lifestyles. Choice is again a factor in the argument. Western lifestyles are often arranged around minimising physical activity – usually replacing it with energy-consuming devices of one kind or another – even to the extent that physical health begins to suffer. Then, in an attempt to attain thermal comfort, we pump up the ambient room temperature, consuming ever-increasing quantities of polluting fuels.

This example illustrates just how complex the question of providing services is. A similar complexity would emerge if we started to look at nutrition. We cannot assume that increasing the consumption of food products leads to increased nutritional health. Of course, it is true that many people in the developing world are undernourished because of a poor supply of food. But in the developed world, the contrary is also true. Over-consumption of food – often of low nutritional value – is endemic in Western industrial economies, resulting in high rates of coronary disease, and compromising individual health.

Providing services is therefore not the same thing as providing material goods. This conclusion emerges so clearly from any discussion of the provision of services, and from any examination of the material intensity of providing them, that it simply cannot be ignored. Clearly, however, it leads us directly into considerations of lifestyle and into realms closer to sociology, psychology and philosophy than we have so far encountered in this book. For the moment, therefore, let us leave these more subjective reflections, and return to the main thrust of the arguments.

In the complex system of the industrial economy, enormous opportunities are open to us to reduce the material intensity with which services of different kinds are provided. Searching for technological opportunities to reduce pollution from industrial processes and extend

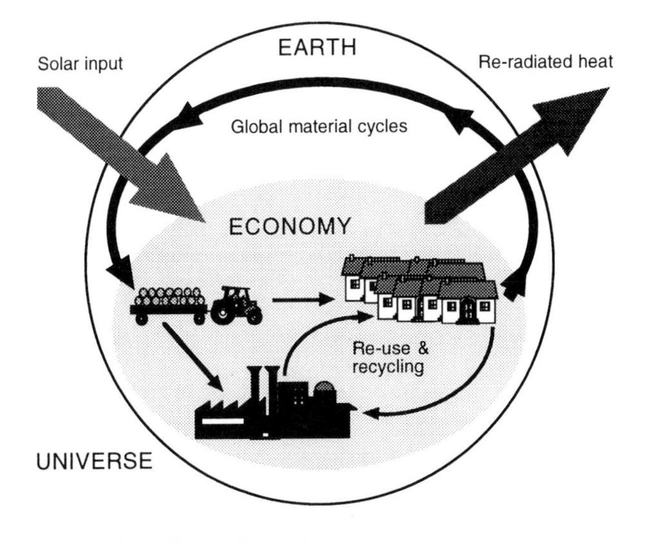

Figure 20 The industrial economy as an ecosystem

product lives is a crucial aspect of the new preventive environmental management strategy. Both of these strategies take us a little closer to what we are trying to achieve: an economic system which is more compatible with the complex ecological system in which it is embedded, a system which aims for high levels of internal materials cycling, and sustains itself within the environment without over-stressing the fragile materials cycles on which it depends for its survival (Figure 20).

5

EASY VIRTUES

Saving money through pollution prevention

TALES OF PROSPERITY

'Pollution prevention pays.' This was the slogan adopted by the US company 3M, one of the first large industrial corporations to take up the new preventive approach to environmental management of process wastes. 3M initiated its '3Ps' programme in 1975 in order to find ways of reducing environmental costs during an economic recession. The experiment paid off. In the first three years, the savings from US facilities alone amounted to over $17 million, with a further $3.5 million coming from subsidiary companies overseas. At the same time, the company eliminated the equivalent of 75,000 tonnes of air pollutants, 1,325 tonnes of water pollutants, 500 million gallons of polluted waste-water and 2,900 tonnes of sludge per year.[1]

This success seems so striking that it is tempting to suppose it was just a one-off accident. Experience shows otherwise. One after another, large companies began to introduce preventive environmental management programmes. Dow Chemicals' WRAP programme (Waste Reduction Always Pays) and the Chevron Corporation's SMART programme (Save Money And Reduce Toxics) were typical of a new wave of environmental initiatives which confirmed that economic savings were to be gained by improving corporate environmental performance.

The experience of the Du Pont company is typical. For years, one of their facilities at Beaumont in the US had been emitting 100 million pounds of waste annually. The Du Pont engineers believed at first that reduction of the pollution would be too expensive. When they took a second look, however, in 1990, they discovered that exactly the opposite was true. The production process could be modified to use less of one particular raw material. As a result, the plant's waste emissions could be cut by two-thirds. Yields went up and costs went down: the action saved $1 million a year.[2]

Despite these successes, scepticism has remained. A recent report from the UK found that almost half of UK companies still have no plans for waste minimisation, and do not even keep track of the costs of generating wastes. One of the arguments has been that such obvious economic advantages as those experienced by 3M and Dow Chemical are available only to large companies. But the evidence no longer bears this out. In 1985, the New York-based non-governmental organisation INFORM documented the impacts of waste reduction programmes at 29 organic chemical manufacturing companies in the US. The selected companies included small, medium and large companies, with a variety of processes, making a variety of products. The study identified 44 waste reduction activities at these 29 plants in 1985 accounting for about 7 million pounds of waste. This only amounted to around 1 per cent of the total waste generated by the 29 plants. But the significant fact uncovered by INFORM was that **every time a plant looked for preventive waste reduction measures, major opportunities were found which saved the plant money**.[3]

A follow-up study on the same 29 plants confirmed these findings. By 1992, a total of 181 reduction activities were reported in the 29 companies, with an average reduction per waste stream of over 70 per cent. These reductions were accompanied by average increases in product yields and average savings of over $350,000 per year. For every dollar which had been spent there was an average saving of $3.50, and the average time taken to pay back the initial capital investment on the measures taken was just over a year.[4]

Nor were these findings limited to the United States. There is now an increasing number of successful projects which have shown that it is possible for industrial companies of all sizes to reduce environmental impact and make money. Amongst them the Swedish Landskrona project, the Dutch PRISMA project, Project Catalyst in the UK, the Aire and Calder Valley project (also in the UK), and the Austrian ECOPROFIT project. Box 3 summarises some of the results from these different projects.

BOX 3 EXAMPLES OF SUCCESSFUL POLLUTION PREVENTION

The following brief summaries describe some of the preliminary results from five regional case studies in pollution prevention. Some of the results are expressed in the form of a payback period. The reader unfamiliar with this terminology is advised to refer to Box 4 (pp. 94–5) for clarification.

THE AIRE AND CALDER VALLEY PROJECT

Established in 1992, to address water pollution problems in the Aire and Calder catchment area in Yorkshire in the UK, the results after the first 18 months of the project were as follows:

- 11 companies took part, including 5 chemicals companies, 2 printing companies, 2 soft drink manufacturers, a laundry and British Rail;
- a total of 542 measures for improving process efficiencies were identified, of which only 2 per cent were later discarded as infeasible;
- the measures resulted in reductions in emissions to rivers and sewers of 285 tonnes per year of chemical oxygen demand and over 300,000 cubic metres per year of suspended solids;
- 10 per cent of measures to reduce waste were cost-neutral; 60 per cent of measures to reduce waste had a payback of less than one year;
- the 11 companies made financial savings of £12 million a year between them; further savings of a similar magnitude were identified for subsequent years.

Source: 'Waste Minimisation – a route to profit and cleaner production', an interim report on the Aire and Calder project, CEST, London, 1994.

THE PRISMA PROJECT

The PRISMA project was established in 1992 as a joint project between the University of Amsterdam and Erasmus University in Rotterdam to investigate the potential for waste minimisation in the two cities. Results of the project at the beginning of 1994 include the following:

- 10 companies took part in the project; these included chemicals companies, printing shops and plating companies;
- a total of 164 waste reduction measures were identified, of which 45 have already been implemented;
- the implemented measures included good housekeeping measures, process modifications and substitutions of input materials;
- of the 45 measures which have been implemented, over 40 per cent were cost-neutral, i.e. they involved no additional capital expenditures; a further 25 per cent of the measures paid back the initial investment in less than a year;
- 7 per cent of the measures which have been implemented were cost-increasing, i.e. they did not have direct economic savings attached to them for the companies concerned.

Source: N. Johnstone 'Cleaner Industry, Cleaner Cities', Centre for the Exploitation of Science and Technology, London, paper to the CityTec '94 conference in Barcelona, February 1994.

PROJECT CATALYST

Project Catalyst was set up in 1993 to promote waste minimisation and improved environmental management among companies contributing to environmental loads in the Mersey Basin in the UK. Results after the first 18 months of the project are:

- 14 companies are taking part in the project, involving a number of different kinds of industries, including chemicals, food processing, electronics and light manufacturing;
- reductions of over 1.8 million cubic metres of polluted waste-water have been identified, along with reductions of 12,000 tonnes of solid waste a year;
- the 14 companies are expected to save a total of £8.9 million a year.

Source: UK Department of Trade and Industry press release, 27 June 1994.

THE LANDSKRONA PROJECT

The Landskrona project was established in the autumn of 1987 by the TEM foundation at the University of Lund in Sweden. The objective was to examine the potential for profitable pollution prevention opportunities in small and medium-sized firms in the Landskrona region in southern Sweden. The project was completed in 1991.

- 6 companies took part in the project, including 3 engineering and manufacturing firms, 2 graphics and printing companies, and 1 chemical company;
- environmental measures included improved accounting and production planning, substitution of alkali degreasing for trichloroethylene degreasing, substitution of powder painting for solvent-based painting; substitution of water-based inks for solvent-based inks; implementation of closed rinse water system, and installation of ultrafiltration equipment.

Source: L. Siljebratt, 1994, 'Pollution Prevention – a profitable investment', Foundation of TEM, University of Lund, Sweden.

THE ECOPROFIT PROJECT

ECOPROFIT = ECOlogical PROject For Integrated environmental Technology

ECOPROFIT is a city-wide pollution prevention initiative in the Austrian city of Graz. Started in April 1991 under the EU's Eureka-PREPARE programme, the results after only one year were:

- in 5 companies (3 printing shops, 1 garage, 1 coffee roastery and chain store group) 54 waste minimisation options were found (24 in the garage, 20 in printing, 10 in the chainstore group);
- 24 per cent of the measures proved profitable in under 1 year; 30 per cent paid back in less than 2 years; and 15 per cent of the measures were neutral in costs;

- at least 50 per cent of the suggested measures therefore made sense economically as well as ecologically and were accepted immediately;
- substantial improvements regarding air emissions and wastes have been achieved: emissions of halogenated hydrocarbons and some toxic wastes could be reduced by 100 per cent; air emissions of volatile organic compounds (VOCs) have been reduced at some applications by 70 to 90 per cent; several substances can be used again through recycling now, and there are waste reductions up to 82 per cent;
- a substantial cost saving has been achieved in the companies.

Source: K. Niederl and H. Schnitzer, 'Greening the Local Economy', paper presented at the Global Forum, Manchester, 1994.

ECONOMIC SCEPTICISM AND OPTIMAL POLLUTION

These umbrella projects and hundreds of other individual experiences corroborate what appears to be a very general trend: preventive environmental protection really does carry with it the potential for economic benefits.

One of the reasons for scepticism amongst industrialists is that this conclusion runs almost completely counter both to the conventional viewpoint and to their own previous experience. The received wisdom has always been that environmental protection costs money. It would be fair to say that this general view has informed much of the political decision-making about environmental matters. Equally importantly, it has informed the reactions of many industrial lobbies, who have sometimes tended to regard environmental protection at best as a necessary liability, and at worst as an unnecessary nuisance.

It is not too difficult to see where this perception has come from. Cleaning up environmental damage caused by industrial emissions *is* very costly. These are the lessons from the incidents – such as the one at Love Canal – cited in Chapter 2. But clean-up technologies – end-of-pipe filters, scrubbers and collectors – are *also* expensive. The economic impact of these 'add-on' technologies is an 'add-on' cost over and above the basic costs of production (Plate 2). For example, the cost of fitting clean-up technology to just one 4,000 MW power station in the UK is expected to reach almost £700 million.[5]

Plate 2 Add-on technology adds on cost – a power station scrubber in West Germany

Source: © The Environmental Picture Library/H. Girardet

This additional investment contributes nothing to revenues, but may involve substantial capital outlays. It might also involve higher running costs (such as the cost of limestone for flue gas desulphurisation) and operational overheads (such as storing and managing substantial quantities of gypsum waste). The more extensive the environmental

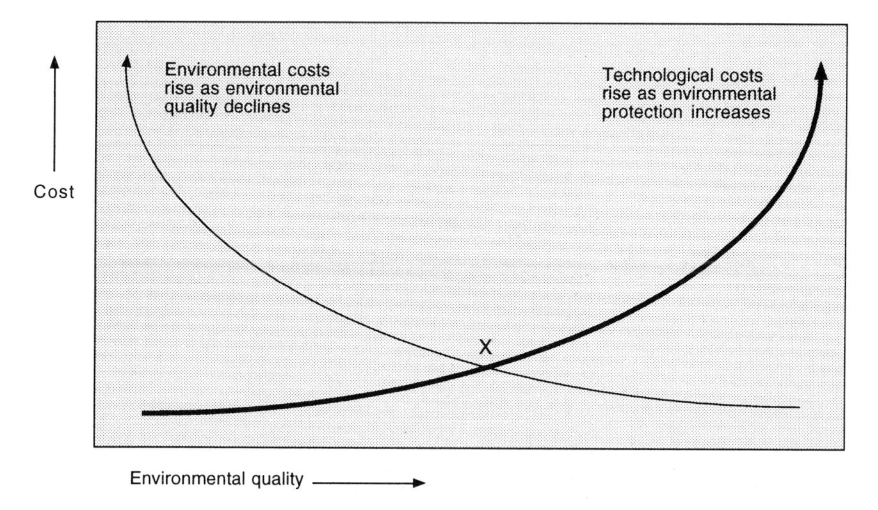

Figure 21 Economics of environmental protection – the traditional view

regulation and the tighter the environmental standards, the greater is the expense. Low-efficiency filters will have to be replaced with higher-efficiency ones as standards improve. Raw material inputs will be in greater demand. Costs will rise accordingly.

This view of the economics of environmental protection is summed up by the graph in Figure 21. The economics of environmental protection is represented by a rising **technology cost curve**. The curve in Figure 21 represents the marginal cost of providing the next unit of environmental protection. According to the conventional view, the costs of environmental protection rise ever more steeply as environmental quality is continually improved.

Based on this picture of rising marginal costs economists have been led to argue that environmental expenditure should be allocated according to 'optimal levels' of pollution. These 'optimal levels' of pollution are determined by the point at which the cost of reducing the next unit of pollution exactly matches the economic benefits of doing so. This economic benefit is determined in its turn by the present and future costs of environmental damage. Figure 21 shows how the theory works. The leftmost curve represents the costs of environmental damage: the costs rise as environmental quality deteriorates. According to the theory, the point X marks the optimal level of pollution. At this point, the cost of extra environmental protection exactly equals the cost of further environmental damage.

Later we shall see that this **theory of optimal pollution** is flawed precisely because it fails to account for the economics of preventive environmental management. But for the moment, I want to stay with the conventional economic view of environmental management. This view – that increased environmental protection means increased costs – has supported a particular attitude towards wealth creation.

The 'trickle-down' theory – which I shall revisit in a later chapter of this book – is really a theory about social equity, but it also has a kind of 'side-kick' theory relating to environmental protection. According to the theory, the top priority for development is to pursue economic growth. The social side of the argument is that this growth will then ensure that everybody in the population achieves a better standard of living, by continually raising the level of wealth in the country. In particular, economic wealth is supposed to 'trickle down' to those who are least well-off, raising the standards of the poor as well as the incomes of the rich. The environmental side of the theory is that economic growth is essential before environmental protection is possible. 'Environmental protection costs money; economic growth provides that money.' That is the way the argument goes.

HOW PREVENTION SAVES MONEY

In this context, the idea that environmental protection can actually save money is clearly a radical one. So it is worth outlining some of the reasons why this might be the case. It is also essential, of course, to try and identify the limitations of such an idea. Can it be true that environmental protection *always* saves money? Or that it saves *everyone* money? Or that the economic savings are immediately accessible and visible? The answer to all of these questions is: probably not. If it were otherwise, we should certainly expect to see far fewer environmental problems than we have seen. So we have to face the possibility that there are some limitations to the idea of economically favourable environmental protection. Before doing so, however, I shall outline some of the reasons why pollution prevention can be expected to pay.

In order to understand fully these economic deliberations it is necessary to know something about the practice of economic cost-benefit analysis. Basically, this is just a method for calculating the costs and benefits associated with a particular capital investment. The difficulty encountered is this: how do we balance out a one-off, upfront, capital cost against a stream of future running costs and benefits? For the benefit of the unfamiliar reader, three different economic balancing techniques are outlined briefly in Box 4. The first criterion – simple payback – is the easiest to understand, but not necessarily the most satisfactory. The other two involve the process of **discounting** capital costs. This complex procedure is related to the idea that interest is charged on the loan of money to borrowers: the **discount rate** in Box 4 is usually set according to the rate of interest a firm would encounter if they had to borrow the capital. Since I will return to the question of charging interest on borrowed money in a later chapter, it may be worth spending some time looking through the procedures outlined in Box 4. But a full understanding is not absolutely essential for what follows.

Let us look now examine the economics of preventive environmental management. We have seen that improving material and energy efficiencies represents one of the central elements of the preventive paradigm. Essentially, this strategy means that fewer material inputs are required to provide the same output. But raw material inputs represent costs for the company. So any strategy which enables the company to reduce these inputs while maintaining their output represents a source of cost savings.

Clearly, there may be some investment costs involved in making these material savings. But the experience of a large number of industrial enterprises indicates that many of these investment costs are very low. Improved housekeeping and stock-taking, better maintenance and repairs, and simple process modifications can all lead to less 'leakage' in the system, and to an overall reduction in the material needs of the company. Some good housekeeping measures and minor process modifications may actually have no initial investment cost at all. Box 5 gives several examples of low-cost measures which have led to raw material cost savings for the companies involved.

Efficiency improvements can also result in reduced running costs, reduced waste management costs, and sometimes lower capital costs. For example, innovative changes to the galvanisation process allowed a metal-processing company in France to reduce capital costs by two-thirds, compared with the traditional process.[6]

BOX 4 ECONOMIC TECHNIQUES FOR INVESTMENT ANALYSIS

A variety of different techniques is used to analyse the economic viability of capital investments. These include the **simple payback period, the net present value** (NPV), and the **internal rate of return** (IRR). These techniques are described in turn below. In each case we shall suppose that the investment costs or capital costs (C) are laid out in the first year of the project. In addition, we will assume that the annual running costs (c) and the annual benefits from the investment (b) are constant for each year of the project. Since we are really interested here in investments which might prove profitable we will suppose that b is greater than c.

SIMPLE PAYBACK PERIOD

This is the simplest of the three methods of determining the economic cost-effectiveness of a particular investment. All that we do is calculate the net annual benefits from the investment by $b - c$. We then divide the capital cost C by these net annual benefits. This tells us the number of years it takes to 'pay back' the initial investment: The simple payback P is given by the formula:

$$P = C/(b - c).$$

A company may use a simple cut-off figure – e.g. a three-year payback – to decide on the financial acceptability of a project. Alternatively, the company may use payback periods to rank different prospective investments and then assign investment capital to those with the best paybacks.

NET PRESENT VALUE

The Net Present Value (NPV) reduces all present and future costs and benefits associated with an investment to a single figure. To do this it uses a procedure known as **discounting**. Briefly, future costs and benefits are taken to have a lower value than present costs and benefits. We can think of the **discount rate** (r) as the rate of return which is required on capital invested by the company.

The higher the discount rate, the lower the value of future costs against present costs. For example, a cost of $200,000 which occurs twenty years in the future has a net present value of $44,000 at 5 per cent and $10,400 at 10 per cent discount rate. The further into the future costs and benefits arise, the lower their value compared with present costs and benefits. For instance, if the $200,000 cost mentioned previously is delayed for a further twenty years (i.e. it occurs 40 years in the future) the net present value becomes $17,400 at 5 per cent and only $1,700 at 10 per cent discount rate. (From these numbers we can see why the costs of decommissioning a nuclear power station – which may occur up to 140 years after the commissioning of the plant – have had so little bearing on decisions to invest in nuclear power.)

The net present value of the overall investment is then calculated by adding up the net present values of each future cost and benefit from the investment for each of the n years over which the equipment remains operational, including the initial capital cost of the investment.

When the annual costs (c) and benefits (b) of the project are the same for each year after an initial capital investment C, the formula for calculating NPV is given by:

$$\text{NPV} = -C + \sum_{(m=1 \text{ to } n)} \frac{(b-c)}{(1+r)^m}$$

A project is supposed to be acceptable (at a given discount rate) if the NPV (at that discount rate) is greater than (or equal to) zero. It is unacceptable (at that rate) if the NPV is less than zero. Alternatively, projects may be prioritised by the investor so that those with higher NPVs are preferred over those with lower NPVs.

INTERNAL RATE OF RETURN AND REQUIRED RATE OF RETURN

The internal rate of return on a project is the discount rate r which makes the NPV exactly equal to zero. The required rate of return is the internal rate of return which a company demands from a project before that project is seen as an acceptable investment. If the actual internal rate of return is less than the required rate of return the investment will be unacceptable; but if the internal rate of return is greater than the required rate of return, then the investment will usually go ahead. Note, however, that a company may also use the internal rate of return to prioritise investments, and allocate investment only to those projects with the highest rates of return.

In other cases, substantial capital investments may be needed to replace one kind of technology with another. For example, the replacement of traditional mercury cell technology in the production of chlorine with membrane technology requires a substantial financial commitment by chemical companies. The robustness (in economic terms) of this kind of investment usually depends on whether the existing equipment is still operational, or whether it is reaching the end of its useful life. In many cases, natural recapitalisation – the investment in new technology to replace old or worn-out technology – will provide substantial opportunities to invest in new, cleaner production technologies (such as the membrane technology). These new technologies can sometimes be less expensive to install than the older technologies. Even when they are more expensive to install than the earlier technology, lower 'running costs' of the new technology can still justify replacement.

The strategy of substitution can also lead to economic savings. Some examples are provided in Box 6. These savings sometimes result from the fact that the substituted materials or processes have lower raw

BOX 5 EXAMPLES OF LOW-COST MODIFICATIONS TO INDUSTRIAL PROCESSES

- A manufacturer of plumbing equipment in the US was producing hazardous waste from an electroplating operation. But a significant proportion of the waste generated came from plating parts which were later found to be defective. A simple procedural modification was made: inspect the parts for faults before plating rather than after plating. Considerable reductions in hazardous waste generation – and raw material inputs – were achieved.
- 30,000 gallons of cyanide-contaminated waste-waters were being generated each year by a US Defense Department metal-plating operation. Much of the contamination was the result of plating solution clinging to the plating parts after they were removed from the bath, and contaminating the rinse water. Drain boards were installed with a capital cost of $900, cutting cyanide wastes by 90 per cent and yielding monthly savings of $784: a payback of just over a month.
- Wastes from a car paint shop in the ECOPROFIT scheme in the city of Graz (see Box 3) arose because of the need to mix fixed quantities of paints for paint jobs of variable size. The development of a computerised mixing system could dispense 40,000 shades of colour in variable quantities and offer an estimated payback of three months.
- Simple design changes to the rinsing system at a Polish plating plant, and the addition of an ion exchange, have reduced the metal-contaminated waste streams by more than 80 per cent. The total capital investment was $36,000. But the annual savings are over $190,000, leading to a payback time of two months.
- An EXXON plant in the INFORM study (see text) achieved a 90 per cent reduction in evaporative losses from chemical storage tanks, simply by installing 'floating roofs' on 16 of the tanks containing the most volatile substances. The modification resulted in savings of $200,000 a year.
- Another plant in the same study made simple operational changes to its rinsing procedures, introducing a two-step process which allowed for recovery of concentrated chemicals at the first step, and reducing the water to dilute second rinse wastes. More than $100,000 were saved in raw material costs on top of the savings in pollution control costs.

Sources: Cleaner Production Worldwide, UNEP, 1993; Hirschhorn and Oldenburg, *Prosperity without Pollution*,1991; INFORM, 1985, *Cutting Chemical Wastes: what 29 organic chemical plants are doing to reduce their hazardous wastes*, by D. Sarokin, W. Muir, C. Miller, S. Sperber, INFORM, New York.

BOX 6 EXAMPLES OF PROFITABLE SUBSTITUTION IN INDUSTRIAL PROCESSES

- A Swedish lighting company, Thorn Järnkonst, substituted the oils which they used to 'cut' aluminium sheets with biodegradable oils. This change allowed them to replace the trichloroethylene degreaser needed to remove the oils with an alkaline degreaser. The alkaline degreaser was cheaper than the trichloroethylene, and did not require the installation of expensive recovery equipment.
- The same company substituted electrostatic powder painting for a solvent-based lacquering process. The investment had a payback period of only 11 months.
- A US printing company, Cleo Wrap, spent six years developing water-based inks to substitute for organic solvent-based inks in one of its printing processes. The investment saved $35,000 a year in hazardous waste disposal costs. As a result of the measure, the company also managed to lower its fire insurance premiums and gain some good publicity.
- A Monsanto plant in Ohio modified a phenol-formaldehyde resin process to produce methylated melamine-formaldehyde resins instead. The substitution reduced hazardous waste generation by 89 per cent, saving the company $57,600 a year.
- An EXXON facility replaced oil as the phenol-absorbing medium in its process with another hydrocarbon. The new hydrocarbon/phenol mixture could be recycled back into the process, whereas the oil/phenol mixture had to be disposed of as hazardous waste. The substitution saved $83,000 a year and eliminated 480,000 pounds of waste.

Sources: *Cleaner Production Worldwide*, UNEP, 1993; Hirschhorn and Oldenburg, *Prosperity without Pollution*, 1991; INFORM, *Cutting Chemical Wastes: what 29 organic chemical plants are doing to reduce their hazardous wastes*, INFORM, New York, 1985.

material input costs. But substitution also generally replaces toxic materials with less toxic ones. This means they are likely to require less care in handling, have smaller environmental impacts and incur fewer environmental penalties. Operating costs will generally be reduced because of the lower safety requirements of the substituted materials.

In many cases, a major area of cost saving is in reduced waste management costs. For example, the disposal of solid wastes carries associated costs – such as landfill 'gate fees'; and waterborne effluents may be subject to sewerage charges. Occasionally, there may be other environmental taxes to pay. These taxes are one of the ways in which governments seek to influence corporate behaviour and improve the environmental performance of industry (see Chapter 8). For instance, an emissions charge levied on each tonne of sulphur emitted into the atmosphere acts as an economic incentive to industry to reduce sulphur emissions.

All these waste management and environmental costs are reduced when measures are taken to minimise the generation of wastes at the source. Sometimes, economic savings will go hand in hand with reduced raw material costs. Improved material efficiency means both fewer material inputs to the process and lower environmental outputs from the process.

Another economic area for potential savings is associated with the cost to industry of complying with environmental regulations. Regulation represents another important policy instrument through which governments can attempt to reduce the environmental impacts of industry. Through an appropriate legislative framework the state can impose specific environmental constraints on various aspects of corporate behaviour. For instance, it can legislate for a limit on allowable emissions of mercury into a local river, or sulphur dioxide into the atmosphere. It can lay down conditions on acceptable waste disposal practices. Or it might ban the use of certain hazardous substances. These environmental regulations are enforced through the legal system with financial or even custodial penalties for failure to comply.

The environmental constraints imposed by these regulations may carry significant compliance costs, especially if the regulations are met by the use of add-on or end-of-pipe technologies. As we have already noted, these technologies generally involve capital investment costs and sometimes higher operating costs as well. But preventive investments – whether aimed at improving material efficiency or substituting away from known hazards – may significantly reduce or even eliminate these compliance costs.

The following example illustrates some of these points. Dow Chemical were faced with environmental regulations to reduce suspended solids in the waste-water emissions from a latex process. An end-of-pipe coagulation unit to remove these solids from the waste-water stream would have resulted in landfilling costs of $70,000 a year. They managed to avoid these costs by implementing an intensive maintenance programme to improve seals and close off leaks, and investing $10,000 in a reservoir to hold and recycle the remaining latex leakages.[7]

The cost savings discussed so far all represent important tangible economic benefits from investing in preventive measures. There may also be some other, less tangible, economic advantages to be gained by reducing emissions into the environment from industrial processes. For instance, better environmental performance is very likely to improve the corporate image in the public eye. This, in its turn, may lead to increased shares in the market and improved profitability for the company concerned. Gaining commercial advantage from these strategies does depend, of course, on a relatively high public awareness about environmental issues. And there also has to be some kind of mechanism for relaying information about corporate environmental performance to the public. But the intuitive reasoning seems to be confirmed by a proliferation of corporate environmental reports, audits and position statements; and by the growing number of commercials which highlight a company's environmental credentials.

In summary, then, there are a number of reasons to suppose that prevention can be cheaper for a company than cure, where environmental management is concerned. End-of-pipe technologies generally add both capital costs and operating costs to a firm's balance sheet. Preventive investment may incur some upfront capital costs, but can generate savings in raw materials, labour costs, environmental charges, and compliance costs. It can also lead to improved corporate image and a greater market share. By way of example, Box 7 provides an illustrative investment appraisal using the cost-benefit methodology discussed in Box 4.

DOUBLE DIVIDENDS: OPTIMAL POLLUTION REVISITED

Generally speaking, then, the economic picture that emerges from the application of preventive measures to industrial processes is significantly different from the economic picture illustrated by Figure 21. The rising cost curve of Figure 21 may be a relatively accurate description of the cost of cleaning up emissions by add-on technologies at the end of the pipe. But preventive measures have a very different

BOX 7 ECONOMICS OF PREVENTION – A WORKED EXAMPLE

Suppose that company X invests in a solvent recovery scheme which costs $100,000 to install and recovers 4,000 gallons of solvent each year which would otherwise have been discarded. Suppose that the cost to the company of purchasing fresh supplies of solvent is $8 per gallon and that the cost of disposing of discarded solvents is $3 per gallon. The operating costs of the system are $4,000 per year.

The total annual savings from the scheme are:

Raw material savings:	$32,000
Disposal costs savings:	$12,000
Less operating costs:	–$4,000
Total annual savings:	$40,000

(a) The simple payback on the investment is 100,000/40,000 = 2.5 years.

(b) If we assume that the solvent recovery equipment has a lifetime of 15 years and that the required rate of return on the capital in the company is 25 per cent, then the Net Present Value of the scheme is given by:

$$\text{NPV} = -\$100{,}000 + \sum_{m=1 \text{ to } 15} \frac{\$40{,}000}{(1.25)^m}$$

$$= \$54{,}371$$

Since this NPV is high, the project should be deemed acceptable at the given rate of return.

(c) The calculated internal rate of return (i.e. the rate of return at which the Net Present Value is zero) is 39.7 per cent. Since this is comfortably above the required rate of return of 25 per cent, we can see again that the project should be regarded favourably as profitable to the company.

economic profile. One simple way of illustrating this new profile is shown in Figure 22.[8]

A series of technology cost curves is shown in Figure 22. The individual cost curves have the same kind of profile as the one in Figure 21: as environmental protection is increased, the costs also increase. But Figure 22 also illustrates another kind of shift. Suppose that we start out at a particular point (point X) on a particular technological trajectory (curve A). Certainly, one way of increasing our degree of environmental protection is to travel further up this cost curve, i.e. implement more of the same technology. And there are additional costs associated with this movement.

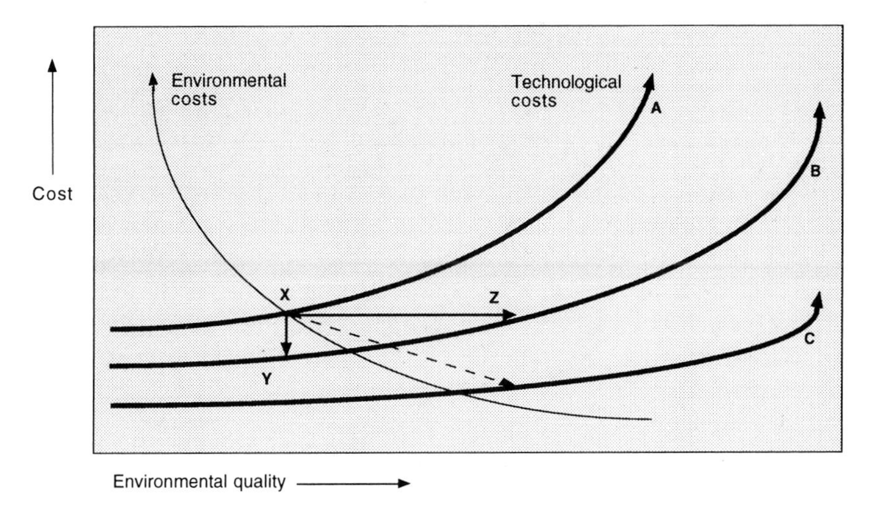

Figure 22 Economics of preventive environmental protection

However, it is clear that a range of other options is available to us if we are prepared to countenance a shift, not along the same technological trajectory, but from one technology cost curve to another. For instance, suppose that we shift to curve B. At point Y on curve B we have the same degree of environmental protection but at a lower cost. At point Z on curve B we have a greater degree of environmental protection for the same cost. Equally, we might then want to look for a further technological shift on to curve C, for instance, where costs are even lower for the same degree of environmental protection.

This is one way of illustrating the difference between conventional approaches to environmental protection and the preventive approach. But we have to be a little bit careful in interpreting this simplistic illustration. It involves a rather broad definition of what we mean by a technology. Essentially, a technology is taken here to be a means of providing a particular service. And it is assumed that the level of service supplied remains constant as we shift between different technological alternatives.

As an example, let us consider the problem of sulphur pollution. Sulphur dioxides are released when coal and oil are burned, because the sulphur content of the fuel is oxidised during the combustion process. Unless measures are taken to prevent it, the sulphur dioxide gas is emitted through the smoke stacks of power stations and factories. When it reaches the atmosphere, the gas is often transported for hundreds of miles before combining with water vapour and falling as acid rain. There is now increasing pressure on countries to reduce acid emissions and firms may find themselves faced with specific regulations limiting the amount of gas they are allowed to emit.

We have already discussed (in Chapter 4) the end-of-pipe response to sulphur pollution: flue gas desulphurisation. This option is quite effective at first in reducing the amount of sulphur emitted. But it imposes additional operating and capital costs on the firm.[9] As levels of sulphur are reduced, the costs for further reductions using end-of-pipe technology rise considerably. For an industrial firm, the preventive strategy would be to consume less fuel by improving the energy efficiency of its processes. Installing more energy-efficient technologies may incur upfront capital costs, but it would save on the company's electricity bills, and reduce all kinds of environmental emissions.

When we come to look at the electricity supply industry, the situation becomes a little more complicated. There are sometimes some efficiency gains which can be made within the electricity production process. For instance, new technologies which gasify coal, and then produce electricity through an integrated combined cycle process tend to be more efficient than conventional coal technology. Generally speaking, though, there are high capital costs associated with these efficiency improvements and the emission reductions are relatively low.

Substituting fossil-fuelled electricity generation with alternative methods of generation is one of the other options which could be considered. In particular, certain **renewable energy technology** options have shown considerable promise over the last few years (Plate 3). These technologies convert natural energy flows (such as wind energy, hydroelectric energy and solar energy) into useful power. The economic costs of these new technologies vary considerably depending on where the technology is located, how strong the ambient energy flows are, and so on. But generally speaking, renewable energy costs have been falling rapidly for a decade.[10] Moreover, they have the advantage not only of eliminating sulphur dioxide pollution but also of reducing or eliminating a number of other air pollutants. Carbon dioxide (which contributes to the problem of global warming) is the most obvious of these other pollutants. Because of their

Plate 3 Solar energy in California – 'the foundation for energy in the twenty-first century'?

Source: © The Environmental Picture Library/James Perez

environmental advantages, the Brundtland Commission has argued that renewable energy technologies should be 'the foundation for energy policy in the 21st century'.[11]

But there is a different way of thinking about pollution prevention from electricity generation. The power station delivers electricity to consumers. As I have already indicated, however, consumers do not really want electricity for its own sake. Rather they want certain kinds of energy services which electricity can provide for them such as light, warmth, cooking heat, mobility and the convenience of electric appliances. By improving the efficiency with which electricity is used in consumer appliances, we could actually reduce the demand for electricity from power stations. The power station would burn less fuel and emit less pollution. This is exactly in line with the aims of preventive environmental management. But it does highlight an extremely important, and rather complex issue. Prevention is not always something that can be carried out within the geographical boundaries of an industrial firm, particularly if that firm is supplying specific commodities (such as electricity) to consumers. This issue is so important that I shall return to it in some detail in the next chapter.

Despite these reservations, the picture shown in Figure 22 allows us to compare the old, received wisdom about the costs of environmental protection, and the new, preventive paradigm. The search for preventive environmental management options becomes the search for new technological cost curves which offer increased environmental protection at lower cost. We should note here that the concept of 'optimal pollution' levels is overturned by the economics of preventive environmental management. Supposedly optimal levels shift continually towards improved environmental quality as we unveil increasingly cleaner technological cost curves.

The industrial experience cited earlier in this chapter and summarised in Boxes 3 to 7 indicates that the search for cleaner technological cost curves is far from fruitless. There is a wide variety of profitable avenues for exploration. Companies from all over the world have benefited from them. And the lesson from this wealth of evidence is perfectly clear: there is significant potential to reduce environmental burdens without compromising economic welfare.

6

PERSISTENT VICES
Understanding resistance to change

INTRODUCTION

If pollution prevention is profitable, as the previous chapter has argued, the obvious question is this: why do we still have environmental problems in an economy dominated by the profit motive? The absence of a profit motive might be one explanation for some of the excessive environmental degradation witnessed in the former Communist bloc of Eastern Europe in the years between 1940 and 1989. But surely, in a market economy, the profit motive should ensure that profitable pollution prevention is implemented? So, why is this not the case? Is it the result of market failure? Or are there some other kinds of obstacles to cost-effective environmental protection? Have we already reached the limit of what it is profitable to do? And if so, where must we now turn if we want to develop a sustainable industrial economy?

Answering these questions is going to lead us deeper into the workings of the industrial economy than we have so far had to travel. In the process of this journey, we are going to discover just how demanding the idea of preventive environmental management is. We must also gain some insight into the complex driving forces of the industrial economy. And only by gaining this insight will we be in a position to go beyond the simple platitude that 'pollution prevention pays'.

Before engaging in this crucial exercise, however, we can dismiss the idea that the potential for cost-effective pollution prevention has been fully exploited. Time and again experience has shown that pollution prevention opportunities remain undiscovered until a firm specifically searches for them. The work of US-based INFORM, and the experience described in the previous chapter from a number of recent local case studies, both demonstrate that there is still a considerable untapped potential for pollution prevention. Armed with this knowledge, it is certainly worth making a determined effort to identify obstacles to the realisation of this potential.

IDENTIFYING OBSTACLES

By way of an example, let us turn once again to the question of energy efficiency. The concept of improving energy efficiency has been around for a long time now. And for over two decades, the message has been more or less the same: there is a very significant potential for measures which improve energy efficiency, reduce energy-related pollution, and are economically cost-effective. As some of this potential has been taken up, more has been discovered. Even so, this considerable potential is still only slowly being exploited.

Clearly it is important to understand the reasons for this slow exploitation of apparently cost-effective energy efficiency. Over the years, therefore, quite a lot of effort has been put into identifying the barriers faced by firms and households in the implementation of energy efficiency.[1]

First, there are barriers to **awareness and information**. At the most basic level, these barriers include the failure by energy consumers even to look for energy efficiency opportunities. This failure sometimes springs from a simple lack of awareness of the technological opportunities available. Sometimes, however, there may be an absence of specific technical expertise. Each technological process has its own characteristics. Devising and implementing technological improvements in complex processes demands both technical skill on the part of engineers and creativity on the part of designers. Occasionally the absence of an appropriate skill and knowledge base leads to a lack of confidence in technological alternatives. Better the devil you know than the one you don't know. The old technology is generally tried and tested. The new one may carry the appearance of risk, particularly to the non-expert.

As technologies change and develop, therefore, continuing investments in research and education are necessary on the part of the firm, if emerging opportunities are to be identified. Equally, there must be an appropriate investment in technical education at the national level if trainee engineers are to have the relevant expertise on environmental performance. Often, these investments can be made without additional costs. Rather it is a question of reorienting training priorities to include appraisal of environmental performance. In the past, however, these matters have either been absent from engineering education or added on, like environmental technologies, only as an afterthought to the mainstream education process.

Where households are concerned, environmental awareness obviously plays a part in motivating consumers. But they also need to be aware that economic savings can be made from energy efficiency investment. In addition, they need access to information and expert advice on new appliances and technologies which might help them to improve their energy efficiency.

These kinds of impediments reside in the knowledge and confidence base of consumers and companies. But there are also **economic obstacles** which appear to strike at the heart of the household's or firm's economic welfare. Technical innovations often require capital outlay. Even though the sums of money involved may not be large, energy efficiency investments nevertheless have to compete with a number of other kinds of demands on company or personal capital. And even when the potential rate of return (see Box 4) on the energy efficiency investment is high, it may still be lower than the rate of return on a competing investment.

Often, the economic obstacle is no more than the absence of an appropriate accounting framework, or any means of identifying a lucrative investment. Individual householders, for example, may often be well placed to reduce their energy consumption by improving insulation, or investing in more efficient appliances. But few of them will have the experience to calculate whether or not the return on a particular efficiency investment is in their best financial interests. Even in small companies, accounting procedures may fall some way short of providing detailed financial assessments of future investments. This problem is compounded by the relatively small proportion of total costs which energy expenditure comprises. For more well-to-do households, the proportion of energy costs to total costs is so small that they probably never figure in investment choices. And for less fortunate households, the energy costs may be a higher proportion, but the absence of investment capital becomes critical.

Finally, the research has identified a number of important **institutional and structural obstacles** to energy efficiency. These obstacles are partly to do with the way in which the energy market has been set up: as a suppliers' market in which a few big companies supply fossil fuels or electricity to a large number of customers. I shall have more to say about the implications of this fact later on. But let me also mention here an important structural problem known as **the tenant–landlord problem**. This term refers to the situation in which it most usually arises. The tenant in a rented property would benefit from the lower fuel costs associated with energy efficiency investment. But he or she is not generally in a position to make capital investments in the property and, in any case, may be moving out of the property before the capital investment is paid off. The landlord, on the other hand, is appropriately placed to make capital investments, but has no incentive to do so because he or she does not pay the fuel bills. Although most obvious in the situation I have described, the same difficulty is to be found whenever there is a **separation of responsibility for costs from the enjoyment of benefits**.

Similar kinds of circumstances are to be found in other fields of preventive environmental management. In fact, many of the obstacles to energy efficiency also appear as impediments to pollution prevention. The lack of awareness and information about pollution prevention opportunities is evident from the number of studies which, once initiated, have immediately identified cost-saving measures.

The question of technical know-how is critical. Process engineers and technical managers usually possess considerable in-depth knowledge about their own particular industrial processes. But they are not always motivated to search for improved environmental performance. This is sometimes because technical education has failed to raise an appropriate awareness of this aspect of engineering. Sometimes it is because they do not have access to information resources which would allow them to identify technical alternatives or design improvements. There is therefore a strong need for information exchanges and clearing-houses in promoting pollution prevention opportunities.[2]

Perhaps more crucially, process engineers and shop-floor managers may not be authorised to make the appropriate financial investments. This aspect of the problem is clearly reminiscent of the tenant–landlord problem in energy efficiency. The tenant–landlord problem refers to a separation of responsibility for costs from the enjoyment of benefits. This is a different kind of separation – the separation of expertise and opportunity from financial responsibility.

Sometimes it is corporate accounting practices which are at fault. The costs of waste disposal and the payment of environmental penalties and charges often remain invisible at the process level. They may be regarded as too small to worry about by the firm's accountants, who do not require managers to identify them on a process-by-process basis. So the responsibility for these costs remains isolated from the responsibility for process management. In these circumstances, there is no incentive for process managers to reduce waste emissions.

At other times, it is the structure of corporate responsibility which raises difficulties. Process managers are often very well aware of the need to reduce environmental emissions, and even able to identify potentially cost-saving investments at the process level. But they may have to compete for funds with other kinds of company investments, in the context of financial decision-making which remains unaware of the potential for improved environmental performance.

Many of these difficulties can be eased by appropriate government policies and committed corporate strategies. I shall have more to say about government policy in Chapter 8. But a very simple procedure which can have broad-ranging impacts is the inauguration of regular waste reduction audits (see Box 1 in Chapter 4) designed specifically to search for pollution prevention opportunities. A related, financial initiative is **full-cost accounting** (Box 8).[3] This procedure aims to ensure that all the relevant costs and benefits (including all waste disposal costs, emissions charges, hidden costs and less easily quantifiable economic factors such as the corporate image) are taken into account in the investment appraisal.

In spite of the value of these simple but important practices, it would be wrong to suggest that all of the impediments to preventive environmental management can be addressed through straightforward procedural changes. There are certainly aspects of preventive environmental management which remain outside the influence of either process managers or company directors. And despite the examples of the previous chapter, there are also some formidable economic imperatives which operate, at times, against the best interests of preventive environmental protection.

BOX 8 ELEMENTS OF FULL-COST ACCOUNTING

CAPITAL COSTS

These are the upfront costs which must be paid for process equipment and hardware. In a full-cost assessment, these costs should also include the following elements:

- an allowance for the installation of utility systems (electricity, water, etc.)
- site preparation costs
- engineering contractors' and consultants' fees
- the costs of start-up training
- the salvage value of displaced equipment

PROCESS OPERATING COSTS

Process-related operating costs are those which are specific to the process or product line in question. These costs are usually most visible to conventional accounting frameworks. But a full-cost assessment should always make sure that the following costs are included:

- cost of raw materials and supplies
- labour costs
- revenues from the sale of products and by-products
- any direct disposal or handling charges paid within the process accounts framework.

OVERHEAD OPERATING COSTS

In addition to direct process-related costs, each process will incur a number of overhead operating costs. Often these overheads remain unaccounted for in conventional frameworks. A total cost assessment should always include:

- waste management costs (haulage, storage, handling, disposal charges and emission fees)
- the costs of regulatory compliance
- the costs of materials handling
- legal costs and penalties
- insurance costs

INDIRECT COSTS

Finally, there are a number of less easily quantifiable costs and benefits which should nevertheless be taken into account. These include:

- estimates of future liabilities for environmental damage
- benefits from improved company image
- benefits from enhanced technological expertise and innovation

In all probability, the 'bottom line' for corporate enterprises is an economic one. If there were always measures which would reduce environmental impact and save a company money for very little or no investment, then nobody would be happier to protect the environment than industry. All the evidence in the previous chapter has shown that there are certainly some circumstances in which this will be true. However, there are certain important circumstances in which it will definitely not be the case. First, there are situations in which the structure of costs and benefits within the firm simply does not reflect an accurate picture of the economic advantages to society of pollution prevention.

ENVIRONMENTAL EXTERNALITIES

Certain kinds of economic costs – such as the cost of environmental damage – fall on the community at large rather than on those whose actions are responsible for incurring them. Because these costs lie outside the accounting framework of the 'polluter' they are often called external costs – or **externalities.** There are plenty of examples of these kinds of externalities: air pollution, water pollution, soil pollution, the loss of amenity value from degraded environments, the depletion of natural resources, the loss of natural habitats and so on.

Some of these costs are quite specific and relatively easy to define. For instance, air pollution has impacts on human health, increasing the number of people with respiratory problems, and increasing the demand on the health service which attempts to treat these problems. To take another example, there is an economic loss involved – as well as an environmental one – when a lake becomes too polluted to use for recreational purposes. Other costs are more nebulous. What is the economic cost, for example, of losing a pristine environment (Plate 6, p. 155)? What value should we place on the loss of agricultural land, the erosion of soil, or the extinction of a rare type of lichen?

A whole new branch of economics, known as **environmental economics**, attempts to bring these problems under its remit. A variety of ingenious proposals for costing environmental externalities has been made. **Contingent valuation** techniques are devised which try to assess people's 'willingness to pay' for environmental amenities by asking them in surveys. Critics have argued that these attempts to value the environment are flawed, partly by the difficulty of actually deriving monetary values, partly because the adoption of such values implies an unwarranted concreteness about the exercise, and partly by the moral implications of subjecting environmental protection to our perceptions of economic value.[4] Although it seems clear that we should *value* other species who share our environment, it is not clear that it is appropriate to place an *economic* value on them. Equally, there are moral difficulties to the practice of assigning monetary values to human life. These difficulties are highlighted by the tendency of economic studies to place a higher value on human life in Western economies than in the developing world. From the point of view of 'willingness to pay' for an increased mortality risk, this practice might be justifiable. From a moral standpoint it is highly dubious.

In spite of the difficulties associated with placing actual economic values on certain kinds of external costs, the main point is clear. There *are* some very real costs associated with industrial pollution. Sooner or later these costs will have to be borne by someone. Some of them will fall upon the taxpayer. Others will be felt in terms of reduced environmental quality and lower amenity value. Some costs will not make themselves felt today at all. Instead they will have to be paid in the future: debts from today's activities borrowed against our children and their children.

The crucial point which I want to make is that these external costs are (by definition) *not* paid by the people who are responsible for incurring them. And the upshot is that it becomes too easy for the polluter to pollute. Industry finds it cheap to use materials and to generate wastes; the consumer finds it cheap to buy material goods and throw them away; governments find it cheap to overexploit mineral resources and pollute the global commons. And all this is because the price of using and consuming materials does not reflect the environmental costs associated with them.

There is an important consequence of this situation for the economics of pollution prevention: **because the polluter does not pay for pollution, there is less economic incentive to reduce that pollution**. Raw material savings from pollution prevention measures will be less than they might have been, because the raw material costs do not reflect the full cost of environmental damage. Emissions charges, disposal fees and environmental penalties are often absent or too low. The combined impact of lower input costs and lower output charges means that there is considerably less economic incentive for the *producer* to invest in preventive measures even when those measures carry net *social* benefits.

The situation is obviously worsened when materials prices are subsidised by government policies of various kinds. Often the price paid by consumers for fuel has been lowered by direct or indirect subsidies to the fuel supply industries. The situation in developing countries – and even the so-called 'economies in transition' such as the former Soviet bloc countries – is often much worse. There are seldom any environmental penalties, material costs are often subsidised and, generally speaking, disposal of all kinds of wastes is regarded as free.

Given these circumstances, it is probably remarkable that there are so many opportunities for pollution prevention which still appear profitable to firms. But equally it is clear that the line between profitable and non-profitable investments is shifted by these failures of the economic system to account for external environmental costs. Ultimately this shift means that *less* pollution prevention is carried out in the industrial economy than is profitable to society.

A SYSTEM CASE STUDY

Next, I want to illustrate some of the complexity associated with preventive environmental management using a particular system case study. This case study starts out by considering a specific pollution problem, related to a single toxic substance. But when we address this problem using the preventive paradigm, we are led rapidly into considering a complex system of material interactions. This example serves therefore to illustrate the peculiar demands which preventive environmental management places on the management of the industrial economy.

The starting point for the analysis is this: how do we reduce environmental pollution from the heavy metal mercury? Mercury (which has the chemical symbol Hg) has already been mentioned in connection with a tragic pollution incident in Japan. Acute mercury poisoning (sometimes designated Minamata disease after the Japanese incident) leads to stomach pains, delirium, coma and death. Low-level mercury poisoning includes a number of effects related to the nervous system. The expression 'mad as a hatter' refers to mercury poisoning amongst hatters who used mercury in the preparation of felt.

In fact, mercury is one of a group of metals[5] which are toxic, persistent and liable to accumulate through the food chain. These properties – toxicity, persistence and liability to bioaccumulate – are all the characteristics (see Chapter 3) which combine to produce the maximum potential for hazard.

Where does mercury come from? There are some natural flows of mercury. For instance, it is naturally present in very small quantities in certain kinds of soils and rocks, and is released through natural processes of weathering. But the main source of mercury that flows through the environment is the economy. There is a number of different sources. Some of them are 'accidental' ones. For instance, mercury is present in very small quantities in coal. When coal is burnt, the metal is released and either gets emitted to the atmosphere or ends up in the combustion ash.

At the same time mercury has proved itself to be a very useful material in the industrial economy. It has been employed for a variety of specific purposes: aside from its former use in felt-making, it has also been used traditionally in dental amalgams; it is used in the chemical industry; it is also used in industrial switches, in thermometers and in batteries.

Each of these different uses carries with it different patterns of release of mercury into the environment. For instance, during the lifetime of a dental patient the mercury used in dental amalgams is bound up chemically and considered relatively safe. But some mercury poisoning episodes have resulted amongst dental technicians. There is also increasing concern about the release of mercury from fillings during cremation.

The biggest single use of mercury in many industrial countries has been in the electrolysis of brine (Figure 23) to produce the chemicals chlorine (Cl) and sodium hydroxide (NaOH). Early 'management' of the mercury wastes generated by the **chloralkali** industry relied on a dilute-and-disperse philosophy. Concern over the accumulation of these wastes in the environment and their return to the human food chain led to the adoption of end-of-pipe removal technologies. The sulphide purification process, for example, precipitates mercury

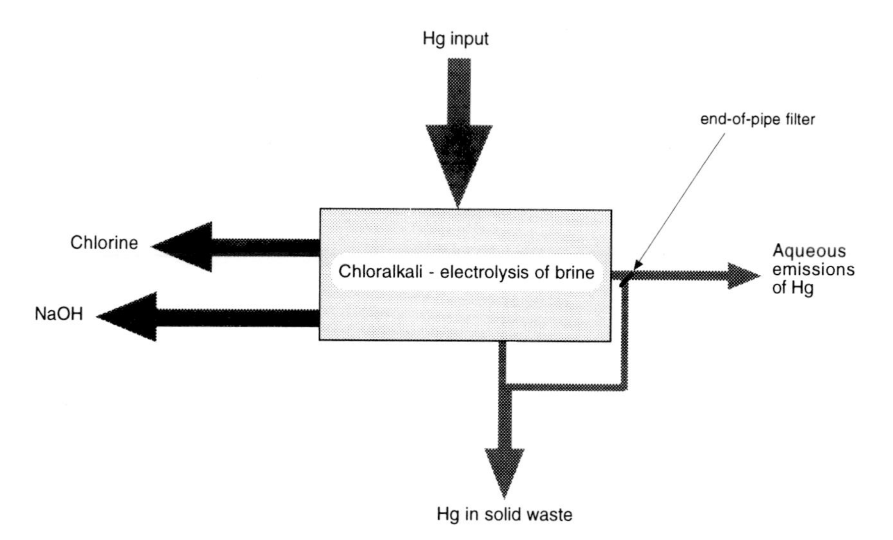

Figure 23 Mercury flows through the chloralkali process

out of industrial effluent as mercury sulphide. The limitations of end-of-pipe processes have already been mentioned. In the case of mercury, a further issue was that the efficiency of the purification process was limited. And there were concerns that remaining levels of discharge into the environment were still too high.

A more preventive solution to the problem of mercury contamination from the chloralkali industry emerged with the recent development of membrane technology. This new technology replaced the need for mercury cells altogether. From the point of view of mercury, this technology therefore represents a completely clean (i.e. mercury-free) alternative. Of course there are still environmental impacts from the new technology. For instance, its energy requirements generate environmental impact. But the development of membrane cells for the chloralkali industry could certainly be described as an important, innovative technological breakthrough.

Most new chloralkali plants now use membrane technology. On the other hand, there is a capital cost associated with implementing a new technology. And in some countries this presents a very significant barrier to environmental improvement. In the UK, ICI has plans to replace its chloralkali capacity entirely with membrane technology. The same is true in Sweden, where it is now law that all mercury cells must be replaced by the year 2010. But a number of older mercury cell plants are still in operation, some of them with no environmental controls operating at all. At one particular facility in the former Czechoslovakia which I visited during the preparatory work for this book, liquid mercury was sitting in pools on the plant floor, and mercury-contaminated sludges and effluent were regularly dumped into an unprotected local landfill site which was believed to be leaking into local ground-water supplies. There was no real prospect of investment in new technologies at the plant.

Even when capital resources are more plentiful than they are at the moment in Central and Eastern Europe, a wise investor would certainly want to question whether reinvestment in process technology was justified. Amongst the factors that he or she would need to assess in answering this question is (see Box 8) *the value of the product or service provided by that process technology*. From a preventive perspective, we would also emphasise the need to look upstream and address the source of environmental problems. And the root of the problem lies beyond the technological process. It is to be found in the demand for particular goods and services.

What are the goods and services provided by the chloralkali industry? In this case, there are of course two products: chlorine and sodium hydroxide, each of which supplies a number of different services. Let us just focus for the moment on chlorine.

Chlorine is itself the subject of environmental concerns. A large number of chlorinated products have environmental implications. Chlorinated organic compounds, for instance, are virtually absent from natural material cycles. Many of them are highly toxic to humans, but because of their organic nature are easily assimilated by biological organisms. Some of them **metabolise** or degrade in the environment in inherently unpredictable ways, and the degradation products are themselves dangerous.

Particular examples of chlorinated organic compounds which have caused environmental concern are: CFCs, used in refrigerators and aerosols and responsible for the destruction of the ozone layer; PCP (penta-chlorinated phenol), used as a preservative in the leather-tanning process but banned in industrial countries because of high levels of contamination in footwear; PCBs (poly-chlorinated biphenyls) used in capacitors and other electrical devices but highly toxic to

Plate 4 Phase out chlorine production? – a Greenpeace protest in Canada
Source: © Heinz Ruckemann, Greenpeace 1990

wildlife, and implicated in soil and water pollution; insecticides like aldrin and DDT which are now banned in industrial countries because of their toxic properties but still widely used in developing countries; and dioxins, which tend to be formed, for instance, during bleaching processes and when chlorinated organic compounds are incinerated.

In fact, there seem to be so many health and environmental problems associated with chlorine and chlorinated substances that some people have argued for the production of chlorine to be phased out altogether (Plate 4).[6] The implications of such a strategy would be quite profound because chlorine is used so widely in the industrial economy. One thing is clear, however: if we are to phase out chlorine use, it makes no sense at all to spend large amounts of money replacing chlorine production capacity. As the economist Herman Daly puts it: 'To do more efficiently that which should not be done in the first place is no cause for rejoicing.' The problem of mercury contamination from the chloralkali industry could be solved more easily by not producing chlorine.

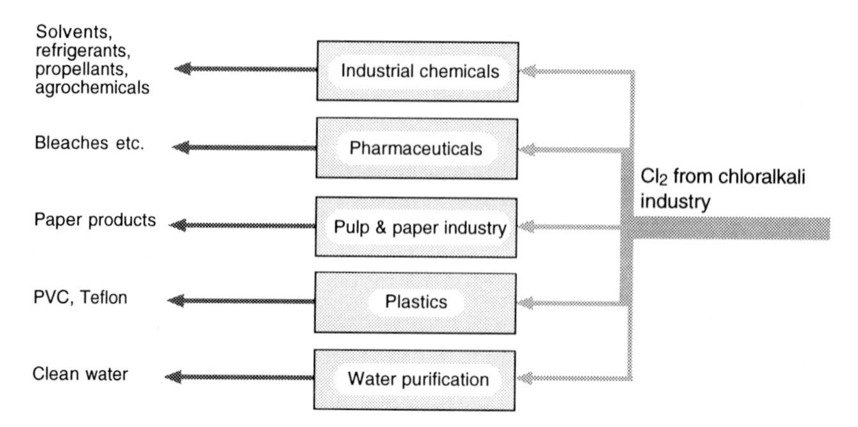

Figure 24 Chlorine as an industrial feedstock

Would it be feasible to phase out chlorine use? What would the implications be for the industrial economy? As you might guess from earlier discussions in this book, it is not so much chlorine itself that we want anyway. Rather we use chlorine and chlorinated products to provide certain kinds of goods and services. In fact, chlorine is such a ubiquitous material in the industrial economy that it is hard to overemphasise the implications of a chlorine ban. As Figure 24 shows, chlorine is widely used as an industrial feedstock in a variety of different sectors. For instance, it is used in the disinfection and sanitation of water supplies, as a bleaching agent in the paper and pulp industry and in domestic products, in the manufacture of plastics, polymers, vinyls, solvents and resins, and in the formulation of a variety of other chemical compounds which have already been mentioned: refrigerants and propellants, insulators, preservatives, and pesticides.

Clearly, some of these uses are more important than others in terms of the service which they provide to society. Phasing out chlorine in one use may not be so drastic in terms of lost service as phasing out chlorine in another use. Alternatives might be found for some uses but not for others. We could only really assess the overall system impacts of a chlorine phase-out by a rather careful analysis on a sector-by-sector basis. Such a level of detail is beyond the scope of this book. But the point is that by looking upstream for solutions to a particular contamination problem we have been led into a complex network of interrelated material flows driven by a variety of different needs and the demand for a number of different services.

As it happens, the problem of reducing emissions of chlorinated organic solvents has been close to the centre of attention in the early pollution prevention initiatives. Because these substances offer both occupational hazards during use and environmental hazards through dispersion and disposal, considerable effort has been put into developing appropriate substitutes for a variety of applications. Substitutes have been developed in the printing industry, for paints and dyes, in the electronics industry and in the chemicals industries.

But organic solvents are only one of the uses to which chlorine is put in the industrial society. Let us look at one of the most important services which chlorine provides – the purification of water supplies. We could not conceive of a chlorine ban without identifying some other suitable means of providing for the decontamination of drinking water. Are there alternatives to chlorine as a water purifier? Can we substitute chlorine-based purification with another process?

In fact, there are other options – the most commonly used alternatives are based on oxygen chemistry rather than chlorine chemistry, and substances such as hydrogen peroxide and sodium percarbonate have been used successfully as bleaches and disinfectants in a number of applications. A different technological route uses ultraviolet light to disinfect and decontaminate water. Before we go into a detailed examination of technological alternatives, however, let us ask another question, a question that takes us even further upstream in the matrix: why do we need water purification services? Could we reduce the *need* for water decontamination by reducing the contamination of water supplies?

This is a complex question because water contamination is a complex issue. Water supplies become contaminated with many different kinds of substances. Pathogens,[7] nutrients, micro-organic chemicals and heavy metals are just a few of the materials which raise health concerns in drinking water. As a water purifier, chlorine is effective mainly in reducing pathogenic contamination – that is, contamination with micro-organisms and bacteria. It is of course extremely important that we do reduce this kind of contamination because these are the kinds of organisms that carry communicable diseases such as cholera. So we might assume that water purification is an irreducible need in this context. But two factors obscure a straightforward conclusion.

First, one of the main sources of this kind of contamination in our water supplies is the practice of using water as a sewerage medium. Raw sewage is a major source of pathogenic contamination. Second, levels of pathogenic activity in water are inversely related to oxygen levels in the water. As oxygen levels fall, pathogenic activity is likely to increase. The addition of oxidants (such as chlorine) into the water is what kills off pathogenic activity. But oxygen levels in rivers, lakes, and ground-waters are dependent on a number of factors. As I pointed out in Chapter 1, excess nutrients can lead to growth in the bacterial and protozoan populations and the subsequent depletion of oxygen. In addition, a number of chemical effluents can have the effect of reducing the levels of available oxygen in the receiving water.

From this perspective we can see that some at least of the need for water purification services (such as might be provided by chlorine) is the result of environmental impacts from other economic activities. If there were less contamination of water supplies from anthropogenic sources, the oxygen levels in the water would be higher, and the natural resistance to pathogenic contamination increased.

If we were to seek preventive alternatives to the use of chlorine for water purification, a number of opportunities present themselves. First, of course, we could reduce our reliance on water as a carrying medium for our sewage. The development of low-water sewerage systems would help. These options are really specific examples of improved efficiency – in this case efficiency in the use of water resources. Quite generally, improvements in water efficiency would reduce the demand we make on water supplies, reduce the contamination that inevitably comes from use, increase the time available for natural purification processes to operate, and increase the available supply from which we could choose clean water. In addition, of course, it is clear that by reducing pollution at the source, both in industry and in the households, we would reduce the burden on our water supplies, reduce the need for chlorination, reduce the need for chlorine production, and reduce the environmental impacts of chlorine production – amongst them, possibly, mercury contamination.

Finally, it would be remiss of me to leave this system analysis without commenting briefly on the second co-product of the chloralkali industry: sodium hydroxide. Caustic soda (as it has been known since before the industrial revolution) has a variety of important industrial uses including the manufacture of soaps, artificial fibres, dyes and paper. Phasing out chloralkali production would obviously raise the question of providing an alternative source of caustic soda with which to supply these uses.

In fact, the chloralkali industry is a relatively recent source of caustic soda in the industrial economy (see Chapter 2). Although the electrolytic process was known in the eighteenth century, it was only after 1890 that sodium hydroxide was actually produced in this way for industrial consumption. Instead, caustic soda was usually produced by treating soda ash (sodium carbonate) with lime (calcium hydroxide). It was not until the 1940s that output from the electrolytic process began to exceed the output from the earlier process. Certainly, therefore, there are possibilities for producing caustic soda which do not involve chlorine as a co-product. A more detailed analysis would need to investigate the resource implications of these alternative processes.[8] But, once again, the more important avenue for exploration, from a preventive viewpoint, is the service output from caustic soda production. This further level of complexity is beyond the scope of this example.

PREVENTION AS AN INSTITUTIONAL CHALLENGE

What emerges from this system case study is that the preventive paradigm entails a very different approach to a particular environmental problem from that implied by the end-of-pipe philosophy. We started out thinking about mercury contamination. We followed the roots of the problem back into a complex network of material flows that embraced chlorine, organic chemicals, nutrients, sewage and water resources (Figure 25). Underlying this complex network, we can identify the need for certain kinds of services, such as clean water, laundry services, dental care and so on. But it is also clear that there are a number of different options for providing those services. And **each way of meeting needs has different material implications and different environmental implications**. This conclusion is clearly reminiscent of the discussion in Chapter 4 concerning the

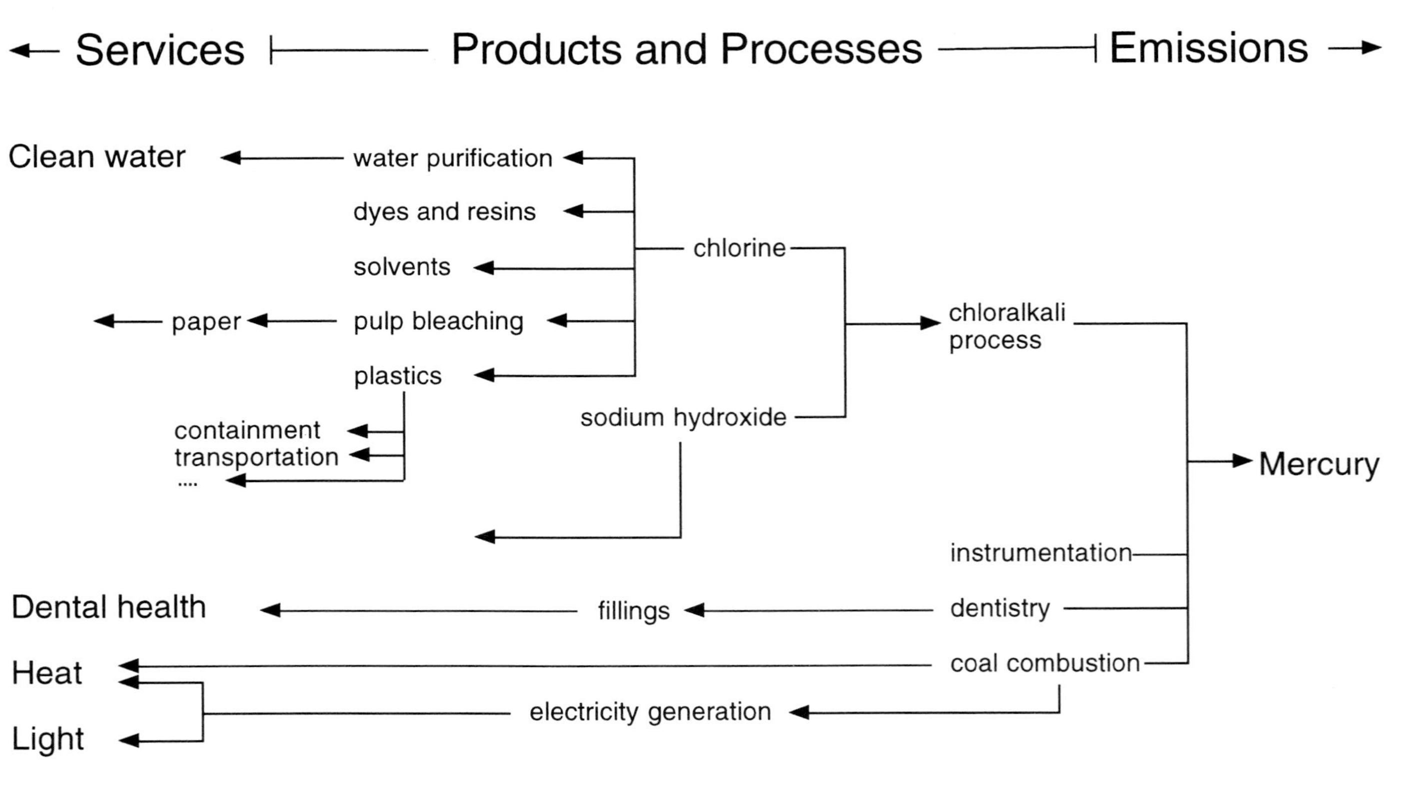

Figure 25 Complex network of material demands and environmental impacts

conceptualisation of services. It is such a crucial aspect of the discussion that I shall return to it in more detail in later chapters.

In the process of this investigation, however, we seem to have left behind the simplicity of a one-dimensional mercury contamination problem. And now we are in a position to see just how fundamental the changes are which the preventive approach demands.

In the oldest scheme of things, environmental management was just a blind faith that nature would solve materials management for us – even when we failed to respect its laws. The intermediate regime required some technological responses. But in institutional terms these technological responses were relatively simple ones. We asked industrialists to clean up their factories, and reduce specific emissions into particular environmental media. Usually, this would just mean adding a piece of technology on to the end of the pipe, to filter out particular contaminants.

Now that we look at the implications of the preventive approach, we find that we are led to intervene at many different places in a highly complex material network. This is no longer something that can be imposed in a unilateral fashion, or undertaken on an individual basis. Rather it demands creativity at many different levels. For instance, it clearly requires a sophisticated physical knowledge base. One aspect of this knowledge base relates to the behaviour of materials released into the environment. But we also need to recognise complex, interactive forces that govern the availability and quality of our material supplies. Equally, technical skills are demanded. The task of providing a safe sewerage system which minimises the impact on water quality is essentially an engineering task; it is not just a question of stresses and strains, but a development of engineering itself as a creative, innovative technical discipline.

This creativity and innovation is crucial throughout the technical basis of the industrial society. But it is not in itself sufficient to meet the demands of the new environmental management paradigm. As this example has illustrated, the preventive approach to a particular pollution problem has implications for many different actors across a wide spectrum, not just for one set of decision-makers in a particular company. We are not simply asking chlorine manufacturers to put a filter on the end of the pipe. We are trying to design a system which reduces the need to produce chlorine in the first place, replaces the use of chlorinated substances with intrinsically less hazardous ones, upgrades our sewage management systems, improves the efficiency of water usage, and protects the quality of our water supplies from industrial contamination.

PROFIT VERSUS PREVENTION?

There is another absolutely critical issue raised by this system case study. The examples of profitable pollution prevention cited in the last chapter indicate that under certain circumstances firms can save money and reduce pollution simultaneously. The discussion of this chapter suggests that **there are commercial situations in which the aims of corporate profitability and the aims of preventive environmental protection are directly at odds with one another**. The reason for this is as follows. Generally speaking, corporate profitability is based on maintaining revenues from the sale of products. But there are a number of important cases in which **the product is the pollutant**.

This is clearly illustrated by the example which we have been discussing. A preventive solution to mercury contamination involves producing and using less mercury, producing and using less chlorine, producing and using fewer chlorine products, and producing and using fewer of the materials which contaminate our water supplies. But this solution suggests a loss of revenue to mercury producers, chlorine producers and producers of a vast range of chlorine products. The same problem arises *whenever* the product is the pollutant. And this is the case for all those companies which are suppliers of raw material inputs: for instance, bulk chemical producers, mining companies, primary metal producers, and fuel supply companies.

In each of these cases, the *consumers* have an economic incentive to reduce their purchases from suppliers. They might do this, for instance, by increasing the efficiency of their processes. Their own corporate environmental performance is improved and costs are reduced. But for the *suppliers* who provide these raw material inputs, the situation may be very different. To them the reduced material demand represents lost revenue. Sales are reduced and eventually the profitability of the company is threatened. For this reason suppliers have a strong incentive to *discourage* reductions in material use, and to find new expanding markets for their products. So we can expect raw materials suppliers to represent a vociferous lobby against the reduction of material flows through the economy.

The economic reality of this position is clearly illustrated by the introduction of state-wide legislation in the US designed to reduce the use of toxics. It is no surprise to find that the first states to introduce this kind of legislation are generally chemicals *users* rather than chemicals *producers*. States with a predominance of chemicals producers have found the introduction of the legislation considerably more difficult, and there has been fierce industrial opposition to it.[9]

The same basic problem affects other kinds of producers. In Chapter 4 I raised the subject of extending product lives. The fewer material goods are produced, the fewer environmental impacts arise before, during and after the useful life. Improved maintenance, reconditioning and the remanufacturing of durable goods may deliver economic benefits to product *users*. But the reduced market for new products which results from this change represents a direct threat to the profitability of the *producer*.

So the implications of the discussion in this chapter extend beyond the welfare of the primary materials producers. They also include the welfare of producers of consumer durables. The important issue here is this: on one side of the equation, products in general pollute; on the other side of the equation, the sale of material products is the fundamental basis for the profitability of many industrial enterprises.

Now we can see that the preventive strategy is not just tinkering at the edges of the industrial economy. Rather it demands fundamental changes to the way in which the economic system operates. This is why I have argued that we need an understanding of the dynamics of the system in which we are operating. In particular, we have to take account of the importance of the profit motive which operates at the heart of the modern industrial economy. In Chapter 2, I discussed briefly the origins of that particular feature of Western civilisation. In intervening chapters I have used the same feature repeatedly as a focal point for examining various strategies for preventive environmental management. Essentially those strategies aim for a substantial reduction in the material intensity of human activities. In the next chapter, I intend to argue that a major reorientation of that same profit motive is absolutely vital to the success of this task.

7

BACK TO THE FUTURE
Reinventing the service economy

INTRODUCTION

'There are three ways of losing your money,' commented the French millionaire Rothschild in the middle of the nineteenth century, 'women, gambling and engineers. The first two are pleasanter, but the last is much the most certain.'[1] The odd thing about this remark is that it was made by a man of undoubted financial acumen at a point in history where, in Britain at least, engineers were about the most profitable investments you could make. It makes sense only when we remember that industrialisation in France lagged half a century or more behind the industrial revolution in Britain. This throwaway comment then becomes an interesting reflection on the structure of pre-industrial economies.

Until the mechanisation of industry overturned things completely, it was not manufacturing but commerce which dominated the thriving market economies of the seventeenth and eighteenth centuries.[2] This domination was reflected in the relative wage structure of merchants and manufacturers: in 1760, the poorest class of merchants in Britain earned as much as the richest class of 'master manufacturers'.[3] And until the eighteenth century, this structural division of the economy was also reflected in the social and political hierarchy.[4]

The industrial revolution changed all that. Mechanisation, the exploitation of mineral resources, the centralisation of production, and massive reorganisation of labour created a new class of rich and powerful industrialists in Britain, and overturned the historical struggle between manufacture and commerce for more than a century.

By the twentieth century, manufacturing was again in relative decline in Britain. Her domination of the world markets for coal, iron and cotton had already been usurped by the expansion of the other industrialising nations: the USA, Germany, France and later Japan in particular. The commercial sector – later known as the service sector – of these powerful, expanding Western economies thrived, as it were, on the back of industrialisation, becoming increasingly more important to the developed world in the second half of the twentieth century. The economic boom of the 1980s (see Figure 8) owed much to the rapid expansion of the service sector. This transition was significant enough to prompt a new description of the industrialised nations as 'service economies'.

These changes in the fortune of different economic sectors are informative not least because they tell us that the industrial economy is a dynamic entity. It is continually changing in sectoral composition, and sometimes undergoes significant reorientations in its profit base. These reorientations have not always led to improved environmental performance. The industrial revolution itself provides ample evidence of that. Nevertheless, the changes themselves indicate that *qualitative* economic development is not only possible but to some extent expected within the existing economic paradigm.

The aim of this chapter is to articulate a particular avenue for qualitative development of the industrial economy. The proposed transition has as its focus the idea of **dematerialisation** of the economy: improvements in material efficiency which reduce both the resource requirements and the environmental impacts of human activities. The emerging vision is of a **new service economy.**[5] This new service economy is not an extension of the 'boom and bust' economy of the 1980s. Neither is it a return to the commercial basis of pre-industrial times. Instead, it is based explicitly on the concept of service which I described in Chapter 4. And it employs many of the specific strategies for preventive environmental management which have been outlined in previous chapters of this book.

FROM GOODS TO SERVICES: REVISING THE BASIS OF PROFITABILITY

From the earliest days of the industrial revolution, economic output has been based on the sale of material goods: first cotton, then iron and steel, leather, metal goods and pottery; later plastics, synthetic textiles, pharmaceuticals, industrial chemicals, mercury, chlorine, and water purifiers. Since profitability depends on economic output, the more of these material commodities a company sells, the better off it becomes. Embedded in the logic of the industrial revolution is the incentive to increase the throughput of materials into the environment.

In Chapter 4 I presented a concept of the economic system which differs in a significant sense from this historical vision. I suggested there that the principal role of the economy was as **a provider of services**. Figure 9 illustrated the basic principles of this role. There are material dimensions to the scheme outlined in Figure 9. In particular, we know from the considerations of thermodynamics that all activities require material inputs and lead to material outputs. But the *functional* output from the system – the one which we intend for some purpose – is not presented in an explicitly material form. This is because the provision of services is not generally representable by material outputs. The units of measurement are different. The concepts are not the same.

Ultimately, we could argue, it is services rather than material products which the economic system ought to be designed to deliver.[6] So the most obvious and yet the most radical revision of the economic system that we could suggest would be to change the basis of profitability of that system from the throughput of products to the provision of services. The suggestion is obvious, because that is precisely what we would like the economic system to do: provide the services which we need to survive and to enjoy our lives. It is radical because it implies a reappraisal of the classic division of the economy between commerce and manufacture, between industry and services.

In the early days, the commercial sector of the economy was mostly concerned with retail trade. Merchants earned their living by buying up cheap material goods in one place – often the colonial markets – and selling them at a profit elsewhere. After small-scale manufacturing developed into a thriving industrial sector, the merchants profited from the rapid throughput of material goods which mechanisation brought with it. And as economies expanded and diversified, the commercial sector became known more broadly as the service sector – that sector of the economy which provided 'services' as distinct from the sector which manufactured products.

The basis of this broad, emerging service sector was really split into three distinct parts. A part of the service sector derived its profitability directly from the retail and trading of material goods. This was the extension into the industrial economy of the formerly powerful base of merchant commerce. Another increasingly important subsector of the service economy was that of financial services. During the economic boom of the 1980s, the fastest rising sector of the British economy was the financial services sector, which practically doubled its contribution to GNP in the decade between 1979 and 1989.[7]

This increasingly important subsector of the service sector appears at first sight to bear little relation to materials throughput, and its rise in industrial economies has contributed significantly to an apparent dematerialisation of those economies. This is an illusion, however. The economic basis for financial services is mostly speculation about commodity prices and trading in investment capital. Speculating on commodity prices both requires and encourages the flow of material commodities somewhere in the economy, and trading in investment capital is empty if capital is not actually invested – generally in materials-based industries. The financial services sector may appear to reduce the materials intensity of the advanced economies. In fact it serves only to promote material throughput and encourage material consumption.

Despite all this, the idea of extending and redefining the service base of the industrial economy is a valid one. To see why, we must look at the third subsector of the service economy. This is the set of economic activities whose profitability is already based specifically on providing certain kinds of physical services: hairdressing, catering, hotels and so on. Naturally, materials and energy are required in order to provide those services. Thermodynamic considerations insist on this. But the crucial point is that these activities do not *base* their profitability on the throughput of materials, and they do not operate under an economic incentive to increase material throughput. In fact, the reverse is true: profitability is based on the provision of a certain service, and there is a built-in incentive to reduce the material throughput associated with that service through the profit motive.

Of course, my suggestion here is not that we build post-industrial economies on the expansion of the hotel trade and hairdressing.[8] These may be the least of our concerns in ensuring national welfare. Rather

I am using this example to point out how a service-based profit motive works in practice. And if it can work for one sector of the economy, why should it not work for another? Historically, commercial activities and manufacturing activities have been divided. But there seems no *a priori* need for this division. And there seems no reason in principle why we should not heal that 200-year-old rift, and reunite manufacturing and commerce under a new conception of the service economy.

The heart of this new conception would be a change in the basis of profitability from the sale of material products to the sale of services. This kind of profitability is already the basis for certain kinds of industries. For a wide variety of other industries it would require significant modifications to existing operations. It would also imply some fundamental revisions of current commercial relationships. What might these revisions mean in practice? In the following sections, a number of explicit examples are presented which illustrate how this kind of change might occur.

SELLING ENERGY SERVICES

First, let us take yet another look at a particular sector which we have already visited several times in this book: the energy sector. Conventionally, the energy sector has been conceived as an industry which supplies fuel to consumers. Accordingly, it tends to be structured around fuel **suppliers** who base their profitability on **maximising the sale of fuels** to consumers. This conception has made fuel supply a very profitable business. But it has also contributed significantly to some of our most serious environmental problems, and acted as a disincentive to investment in improved energy efficiency.

Perversely, this sector is not even structured in such a way as to provide what people want. As I have suggested already, people do not want fuels for the sake of having fuels. Oil is no good to anyone, except in so far as it provides certain **energy services** such as thermal comfort and mobility. What seems to have happened with the energy industry, as with so many other industries, is that the force which drives corporate behaviour has somehow become misaligned with the needs of the individuals for whom we would ideally wish the system to provide.

What can be done about this? In the energy sector, this mismatch between the interests of suppliers and the interests of consumers has been the subject of considerable analysis over the last decade or so. Discussions have mainly focused on the best way to overcome the obstacles which stand in the way of improved energy efficiency (see Chapter 6).

The principal difficulty can be explained in the following way. Suppose that an electricity supply company is looking at the best way to meet the demand for electricity from its customers. One way of meeting a projected increase in demand would be to build a new generating station. Another way would be to reduce the demand for electricity by investing in more efficient electrical appliances in the home. Both ways would deliver the same level of energy services. But the second way would reduce the demand for fuel consumption associated with electricity generation. It is, from the national perspective, a more efficient means of providing the same service; it is also less environmentally damaging; and furthermore it turns out to be much cheaper to improve efficiency in the home than to invest in the new power station.[9]

But there is a problem. The electricity utility is responsible for building power stations, and even has an economic interest in doing so, provided that it can recoup its costs by selling electricity. But individual consumers are responsible for buying and installing their own electrical appliances. The utility has access to the capital needed to build power stations, and can profit from doing so. With the same amount of capital, consumers could buy all the efficient appliances they need, and still have money to spare. But, generally speaking, individuals do not have access to this sort of capital without incurring expensive debts. The interest rates on personal debt tend to be very much higher than those levied on the large utility investor.

The obvious solution to this dilemma – but one which has involved long complicated discussions and considerable wrangling[10] and which even now is not widespread – is for the utility to invest in energy-efficient appliances for their customers, and for the profits of investing in a more cost-effective provision of energy services to be split between the utility and the customer. Under the new arrangement, the price of electricity rises to compensate the utility for its investment. But the number of units of electricity used by consumers falls. Both the utility and the consumers benefit economically from the more efficient provision of services. The same basic logic can be applied to other energy services: utility investments in thermal insulation, and in improved conversion efficiency in the home, in offices, and even in industry.

One of the reasons why this solution has been so difficult to arrive at is that it involves a fundamental revision of the basis of profitability of the utilities. Instead of basing their profits on revenues from the supply of electricity, they would now base some of their profits on investing in what is called **demand-side management:** reducing the demand for electricity by improving conversion efficiencies in the home and the workplace. Quite apart from the change in corporate thinking which this reorientation demands, institutional constraints have also prevented a smooth transition.

For example, many countries impose price regulations which govern rises in the unit costs of electricity. The structure of these regulations has tended to restrict the ability of the utilities to recover investments in demand-side measures. It has proved necessary to renegotiate those kinds of regulations to allow price increases which serve to reduce overall costs. But these difficulties are not insuperable given appropriate institutional will.

It is clear, however, that new arrangements imply new and different commercial relationships in the energy market. Under the conventional conception of the electricity supply industry, the trading arrangement between electricity utilities and customers is relatively straightforward. The utilities produce and distribute electricity to sell to their customers. The profitability of the utilities rests squarely on revenues from electricity sales to consumers. Under the new arrangements, it needs to be possible for utilities to raise revenue, in part, by investing in energy efficiency measures for participating customers. This means that the utilities now become **energy service companies** rather than electricity suppliers. The Sacramento Municipal Utility District in California is an example of the successful emergence of such an energy service company, selling energy efficiency to its customers as well as supplying fuels.

The service concept also presents opportunities for 'third-party' energy service companies – separate from consumers and from the utilities, but engaging commercially with both. One of the main

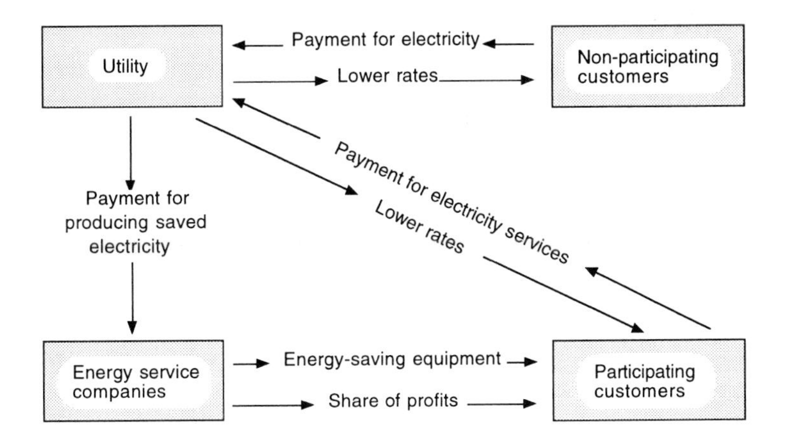

Figure 26 Benefit flows from third-party energy service companies

advantages of these companies is to provide the independent expertise in energy efficiency which is not necessarily inherent in the energy supply companies. Figure 26 illustrates how the revised commercial arrangements might operate.[11]

OPERATIONAL LEASING OF CONSUMER DURABLES

Can we apply the same operational changes to other kinds of commercial activities? In principle, there is no reason at all why not. Let us take an example from a completely different realm of economic goods: consumer durables.

Supplying computer hardware – as usually envisaged – relies on a chain of economic transactions relating to material goods, each step of which provides the profit basis for suppliers. Manufacturers supply computers to wholesalers who supply them (usually) to retailers who supply them to individual consumers. Each step of the chain relies on the profit available from a steady (and increasing) throughput of material hardware. But customers do not particularly want or need an increasing flow of computers. Rather they are concerned with acquiring certain services which computers provide. Recognising this need, some companies have made preliminary attempts to restructure their operations. For example, the development of a 'portable operating system' for one of their client companies by Siemens-Albis follows these ideas.[12] The technical innovation around which this development was based was a decoupling of the software and hardware components of computing systems. The commercial innovation was to introduce producer responsibility for the hardware components throughout the product life.

The idea of this kind of restructuring then revolves around the idea that computer hardware is **leased** rather than sold to users. Rather than buying computers, customers buy **computing services** from their suppliers. And these services include maintenance arrangements, upgrade facilities, and supplier responsibility to take back the computer at the end of its service life.

This may seem like a minor modification to the existing system. But it has potentially far-reaching implications.[13] First of all, when material goods are only leased by the final user, responsibility for them resides with the producer or supplier. This has profound environmental implications. In the existing system, there is no incentive on the producer to design his products for minimal material throughput. On the contrary, there is an incentive to minimise the life of products, so as to prevent the market from saturating, and this will tend to increase the overall material throughput. Under the new arrangements, however, maintenance, upgrade, reconditioning, reuse, and the recycling or disposal of material components would all remain the duty of the supplier. This would transform the incentive towards early obsolescence into an incentive for product durability, material efficiency and environmental responsibility.

The same philosophy is applicable to a number of other durable consumer goods. The long-term, flexible leasing arrangements for long-life photocopiers – first commercialised by Agfa-Gevaert – provides another example of this kind of innovation.[14] The rental of television sets (and now computers) by the British company Radio Rentals is a further example.

Another consumer durable which has some kind of track record in the sale of services is the automobile. One strategy, for instance, reconceives car producers as service agents who 'make their money not primarily by making and selling new cars, but by selling spares, repair and aftercare to keep their own products on the road for a long time'.[15] In fact, this strategy is not so far from a reality in some places. In Sweden, for example, the useful life of cars has been extended considerably in recent years and vehicle life expectancy is now in the region of seventeen years. This extension of product life has been supported by strong marketing of second-hand cars, including high-profile display areas and – in the case of Volvo – used car brochures.

At the same time, this particular example illustrates that the strategy of operational leasing needs to be developed with some care. One of the problems is that product life extension – inappropriately conceived – can stand in the way of technological upgrading, which might in turn lead to environmental improvement. Technological developments in the automotive industry, for example, have placed an increasing emphasis on improved fuel efficiency. This leads to lower fuel consumption on the road, and reduced environmental burdens as a result.

The future of the automotive industry certainly holds further significant changes in design specification and material requirements. Ultra-efficient cars, based on the same light but extremely durable carbon-based polymers from which racing cars are already made, may soon replace existing designs. But the introduction of these newer and more efficient vehicles may be slowed considerably by strategies to prolong the life of existing models. These considerations highlight the importance of designing products capable of future upgrading to improve performance.

But a more worrying problem presents itself when we take a wider perspective on the provision of transportation services. It was apparent from the discussion at the end of Chapter 4 that the provision of services is not the same thing as the provision of material goods. The services themselves are not measured in terms of tonnage of material throughput. And the conception of the service itself is complicated by a variety of different factors. In the final chapter of the book I shall relate this difficulty to the intricate question of satisfying human needs. But the following section highlights some at least of the complexity involved in reconceiving consumer services.

RECONCEIVING TRANSPORT SERVICES

Let us examine again the example of motor transport which was discussed above. From the conceptual viewpoint of the new service economy, we need to ask: what is the service provided by cars? I have already remarked that cars have come to represent a number of value-laden aspects of modern industrial society.[16] They are associated, for example, with concepts of personal freedom. And they also express elements of status and personal taste. But for the moment I want to leave these more subjective matters to one side and concentrate on the primary objective function provided by car ownership. In particular, of course, cars represent one of the material requirements in the provision of a specific service: namely, transportation.

One way of expressing the delivery of services from passenger transportation is in terms of distance travelled per year. We can either count up the total **passenger-kilometres** travelled each year, or we can use the average distance travelled by each person in a year as an indication of the delivery of transportation services. Figure 27 illustrates the delivery of per capita transportation services (excluding air travel) in Sweden from 1950 to 1990.

Several important lessons emerge from this graph. First, the figure shows that car travel now represents a high proportion of the total delivered transportation services. In 1990, around three-quarters of total passenger-kilometres were provided by cars. Just over 20 per cent were provided from public transport (buses, trains and trams). Less than 10 per cent were covered by walking and cycling. This contrasts sharply with the situation in 1950 when around half of all transportation services were provided by public transport. At that time,

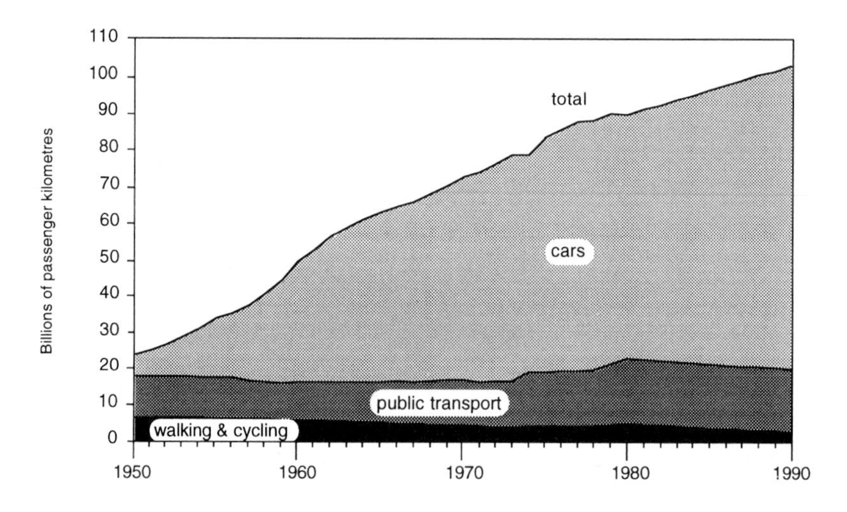

Figure 27 Passenger transportation services in Sweden, 1950–90

Plate 5 Is the car sustainable? – traffic congestion in Mexico City
Source: © The Environmental Picture Library/Matt Sampson

cars provided less than a third of all transportation, and less in total than was provided through walking and cycling.

The material implications of these trends emerge from the realisation that car transportation is considerably more intensive of energy and materials use (per passenger-kilometre) than public transport is. Both are more intensive than walking or cycling. This means that the trend towards increased car use has had very significant impacts on the material intensity of transportation services. But it also raises the possibility that we could reconceive transportation services in such a way as to reduce the material intensity again, by encouraging a shift from cars towards public transport. Many countries are now beginning to do just this – particularly in the towns and cities – where fuel efficiencies are lower, and traffic congestion from cars is an increasing problem (Plate 5). In Sweden, for example, there are a number of specific programmes and incentive schemes designed to encourage public transport over cars.

For long-distance journeys one of the advantages which has been offered by the car is flexibility. Cars allow (in principle) for individual choice over the route travelled, the time taken for the journey, stopover points, and so on. In fact, many of these advantages are being eroded by increasing congestion. But the successful implementation of an alternative service must at least attempt to provide a similar flexibility. In practice, however, the rise of road transport has eroded flexibility in the rail networks. Smaller branch lines have been subject to closure and passengers are sometimes penalised for delaying or interrupting journeys. A comprehensive, flexible and yet materially efficient transport system therefore requires both technological and commercial inspiration. Flexible ticketing is as important as infrastructure investments.

Again, the emergence of a new service network relies on new commercial initiatives. And the idea of leasing transport services rather than selling cars (and fuel) offers some exciting possibilities – some of them novel, and some which have been around for years. The extension of the idea of 'rover' tickets – once a common feature of both the railways and the buses – would be one such possibility. Another would be the development of **travel service companies** which combine long-distance public transportation with short-distance taxi services or car rental.[17] Again, it is clear that these developments imply different commercial relationships from those which exist at the moment. But they also offer considerable potential to reduce the material intensity of transportation services.

What appears once more from these considerations is that the provision of transportation services is not the same thing as the provision of material commodities (cars and fuel, for instance). Conceiving and designing a materially efficient transport service implies thinking and acting strategically in a changing commercial context. In particular, it should be obvious that investment in improved car designs and durability should not overlook – or impede – the development of more efficient ways of providing transportation services.

Before we leave this case study, there is another aspect which is worth commenting on. It is clear from Figure 27 that the *demand* for transportation services has increased dramatically, even on a per capita basis, since 1950. This demand increase has important technical repercussions for the kind of transport system we develop. For example, it is clear that walking and cycling could not provide the same proportion of the total demand in 1990 as they provided in 1950. It simply would not be possible for people to travel the relevant distances on foot or on bike without spending their whole lives doing it. In other words, automotive technology is delivering a *level* of service which is over and above what would have been possible without it. It is clear that we are demanding much more from transportation services now than we did forty years ago.

This raises the question of whether we should accept these demand increases as the starting point for providing services or whether it is legitimate to include **demand-side management** within our remit in reconceiving the transport system. Do increases in passenger kilometres represent real increases in our standard of living? Or are they necessary only to offset new patterns of working, shopping and living? Some at least of the increased demand falls into the latter category. For instance, people generally travel further to get to work than they used to do. It seems fair to assume that the increased commuting distances represent a decrease in welfare rather than an increase.

What is being suggested here is that transportation services are not, in fact, an end in themselves. Rather there are some *underlying* services – such as 'getting to work' or 'going on holiday' or 'buying the groceries' – which transportation provides for us. Once again, we are forced to recognise how complex the reconception of services is. In the case of transportation services, it is certainly not a straightforward matter of providing vehicles and fuels. It is not even a question of delivering passenger-kilometres, although that is an improvement over selling cars. Instead it requires an understanding of the demands placed on the transport networks and the design of appropriate frameworks to deliver the relevant services. Supplying these underlying services is not just a task for road-builders or car manufacturers. It concerns a wide range of actors and decision-makers. In particular, of course, it involves the entire system of *planning* under which new shopping developments, work complexes and residential estates are conceived.

These considerations reveal that this revision of the concept of transportation services is a challenging task, technically, commercially and institutionally. But it offers us the possibility of designing systems which deliver the same level of service with considerably lower environmental impacts than existing systems.

SOME FURTHER EXAMPLES

There are a number of other examples illustrating the way in which services can be provided with reduced material intensity. For instance, let us consider the case of washing machines. Obviously, as a consumer durable, the washing machine might be considered a prime candidate for operational leasing, as described earlier in this chapter. Essentially, service companies would provide, maintain and recondition the machines, and ultimately be responsible for recycling or disposal of the component materials. At the same time, it is legitimate to ask whether or not there is a way of delivering the service (clean clothes) currently provided by domestic washing machines – but with even lower material intensity. As Figure 28 illustrates, the major part of the environmental impact associated with the life-cycle of a washing machine occurs during the use of the machine.[18] In these circumstances, it makes no sense to focus exclusively on minimising waste from the production or disposal of the machines themselves.

Some of the utilisation impacts arise from the energy requirements of washing machines. Others arise from the disposal of soiled water

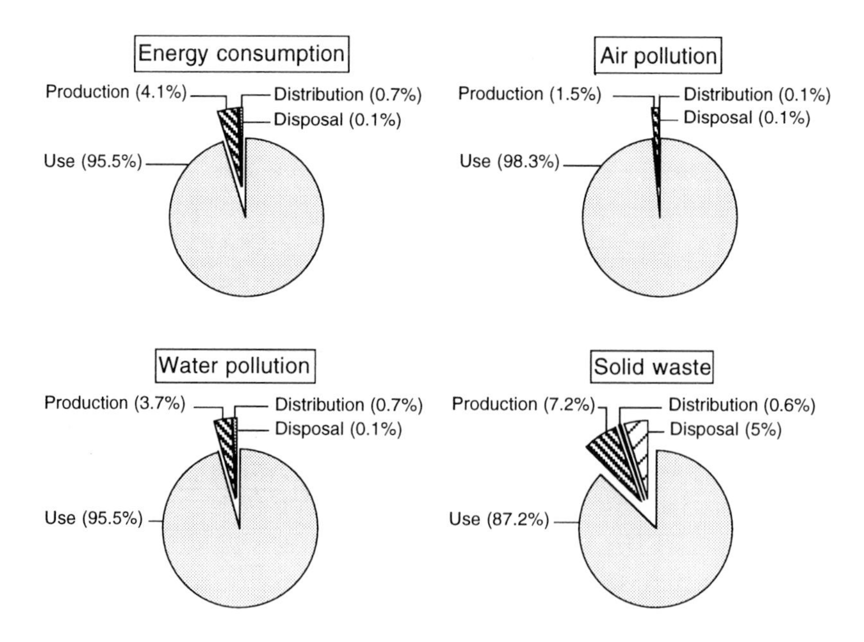

Figure 28 Distribution of environmental loads during the life-cycle of a washing machine

from the washing process. This is clearly one of those situations (identified in Chapter 1) in which certain products (in this case detergents) are routinely dissipated into the environment during use. Several options to reduce the environmental impacts of providing clean clothes present themselves for consideration.[19] In the case of dry-cleaning operations, the solvents which are used to clean materials can be separated from the removed soil and recovered for future use – for instance by evaporation and condensation. If the same idea of separating the soil from the removal medium were possible for wet washing – for instance through membrane technology – the hot water and detergent could be recovered, replenished and returned to the washing cycle. The process of replenishment for further use would involve some commercial innovation, however, requiring an interactive involvement between washing-machine suppliers (ideally operating under a leasing arrangement) and detergent suppliers.

Once the soil stream is concentrated there emerges the possibility of recovering the energy content of the soil – for instance, from recovered animal fats. This option would become more viable in the case of a collective washing facility operating within a local community. Again, the possibility of new commercial arrangements is revealed. Clean clothes might be provided by the sale of local laundry services rather than by the sale of washing machines and detergents. The focus of all these potential developments is on the effective provision of service with the minimum material requirements and environmental impacts.

Another sector which gives rise to routine dispersal of chemicals into the environment is the supply of agricultural chemicals. Historically, the chemicals industry has simply fulfilled the role of *supplier* of agrochemicals: pesticides and fertilisers. Lacking detailed expertise, appropriate scientific knowledge and sophisticated application techniques, farmers have tended to use chemicals widely, inappropriately and in excess. In fact, it has been estimated that the application of herbicides used to control wild oats has traditionally been carried out at rates which are around 1 billion times greater than is necessary to kill the unwanted strains.[20]

Of course, this profligate use of materials is of financial benefit to the suppliers, but it imposes high costs on the farmer as well as posing a number of environmental concerns. As pesticides come under critical environmental appraisal, there have been attempts to reduce these excessive application rates. For instance the amount of material used per hectare has fallen by about 32 per cent since 1984 as a result of the promotion of more responsible use of pesticides.[21] Increasingly the focus of pesticide strategies will be on new application technologies. From the point of view of chemical suppliers, these changes will mean that fewer pesticides are sold. But there are now emerging markets for the chemical companies in the field of **selective application techniques** and **targeted dosage**.

Again, these markets amount to a shift in the basis of profitability. Output is determined by a new functional unit of service delivered – for instance, the degree of protection from unwanted crop strains – rather than by the material quantity of chemicals sold. In addition, of course, I should mention here that there are horticultural aspects to the problem of pest control in agriculture. The choice and application of different farming practices (crop rotation, tilling and so on) can also play a substantial part in reducing the demand for chemicals. The emergence of organic farming methods and permaculture represents another avenue for the pursuit of reduced material intensity in food production.

TOWARDS MATERIALS LEASING

Yet another instance of commercial innovation in the provision of services is provided by the case of the organic solvent kerosene. Used in a number of applications as a degreasing agent, kerosene has been the object of successful commercial 'materials leasing' by the chemicals company Safety-Kleen in the UK for a number of years.[22] Kerosene is supplied to customers on a use-and-return basis. The degraded solvent is returned to the supplier after use. It is then reprocessed and subsequently supplied to the same or another company for reuse. Soiled residues from the reprocessing operation also find use as commercial fuels – for instance in cement kilns. This is an example of the **cascading** of raw materials through a variety of uses as discussed in Chapter 4. Again, the important point to note is that a new commercial arrangement is implied between supplier and user of the material. Effectively, this arrangement sells **degreasing services** to the user.

In an expanded system of materials leasing, we would almost certainly find new structural relationships between commercial agents. In direct analogy to the energy sector, we might find that companies which had traditionally been primary materials suppliers now diversified their activities. Instead of investing purely in supply, they would now have an incentive to invest in improved efficiency by materials users. For instance, in the case of kerosene degreasing, there might now be an economic incentive for the kerosene supplier to invest in solvent recovery in the degreasing plant. Equally, companies which had traditionally been involved in manufacturing consumer durables might choose to invest preferentially in service infrastructure rather than in product supply.

Arriving at this challenging and sophisticated form of dematerialisation would almost certainly rely on radically revised pricing policies, new regulatory initiatives and different 'ownership' structures and liability frameworks. But in principle we could foresee a system in which wealth was not dependent on the profligate trade of material goods. There might no longer be a linear chain comprising materials suppliers, materials processors and materials consumers. Instead, new service-based corporate structures might emerge in which materials management was a much less linear, but much more integrated, process.

8

NEGOTIATING CHANGE
Dematerialisation and the profit motive

INTRODUCTION

At various places in this book, I have dwelt on the underlying profit motive of the industrial economy. Clearly, the pursuit of profit has been a crucial factor in the development of that economy (see Chapter 2). It is also a continuing motivation for improvements in process efficiencies in industry. The examples in Chapter 5 indicate that improved environmental protection and the pursuit of profit are not necessarily natural enemies. But Chapter 6 has revealed that there are certain commercial situations in which profit and dematerialisation are in direct opposition. In fact, this is generally the case whenever the product is the pollutant. Since all *material* products are potential pollutants at some stage during their life, this has led us (Chapter 7) to a revision of the basis of profitability in the industrial economy.

Under the new service economy profitability is no longer based on material products. Rather it is contingent on the provision of services. Although these services inevitably require material inputs and outputs, the incentive to increase material throughput is transformed – via the commercial innovations of the service economy – into a continuing drive for material efficiency.

These considerations reveal that the relationship between the profit motive and the material basis of the economy is intricate and multifaceted. Nor have we exhausted the complexities of the situation. In particular, there are three important questions which still need to be addressed. In the first place, can we rely on the profit motive itself to transform the economy for us in the way described in the previous chapter? Actually, it is relatively straightforward to answer this question negatively. But having done so, we then need to ask what else we should be doing to accomplish the necessary changes. And having answered that question we are still left with one critical issue.

The profit motive is the basic driving force behind economic growth. As the economy expands, activity levels increase. The increase in activity levels means an increase in material throughput, *unless* we can improve material efficiency fast enough. In environmental terms we could say that the profit motive has turned out to be a distinctly double-edged sword. On the one hand, it provides the motivation for improved material efficiency. On the other hand, profit on investment has provided the impetus for increased economic output.

In the light of these remarks, the most important question to be addressed in this chapter is whether dematerialisation can outpace – and continue to outpace – economic growth. If this could happen then there is absolutely no irresolvable conflict between the profit motive and the strategy of dematerialisation. If it cannot happen, then we may well be forced to face the possibility that there are environmental limits to economic growth. The implications of this position for conventional notions of development are even more profound than the prospect of revising the basis of profitability. But before we deal with this more complex issue, let us return to the question of whether the industrial economy will naturally evolve towards dematerialisation, under the impetus of the profit motive.

DEMATERIALISATION: IS IT HAPPENING ANYWAY?

Even from very early on, industrialists were prompted to improve the technological efficiency of their processes for economic reasons. For example, between 1791 and 1830, the volume of coal consumed to produce one tonne of iron was reduced by over 50 per cent, from just over 8 tonnes to just over 3.5 tonnes.[1] The examples cited in Chapter 5 indicate the considerable potential for profitable efficiency improvements which still exist today. The pursuit of profit provides some if not all of the motivation for these improvements.

Quite generally, it has been argued that profit-driven industrial development is accompanied by a *natural* tendency towards dematerialisation. This idea was first proposed by Malenbaum who introduced the concept of **intensity of use** to measure the quantity of a particular

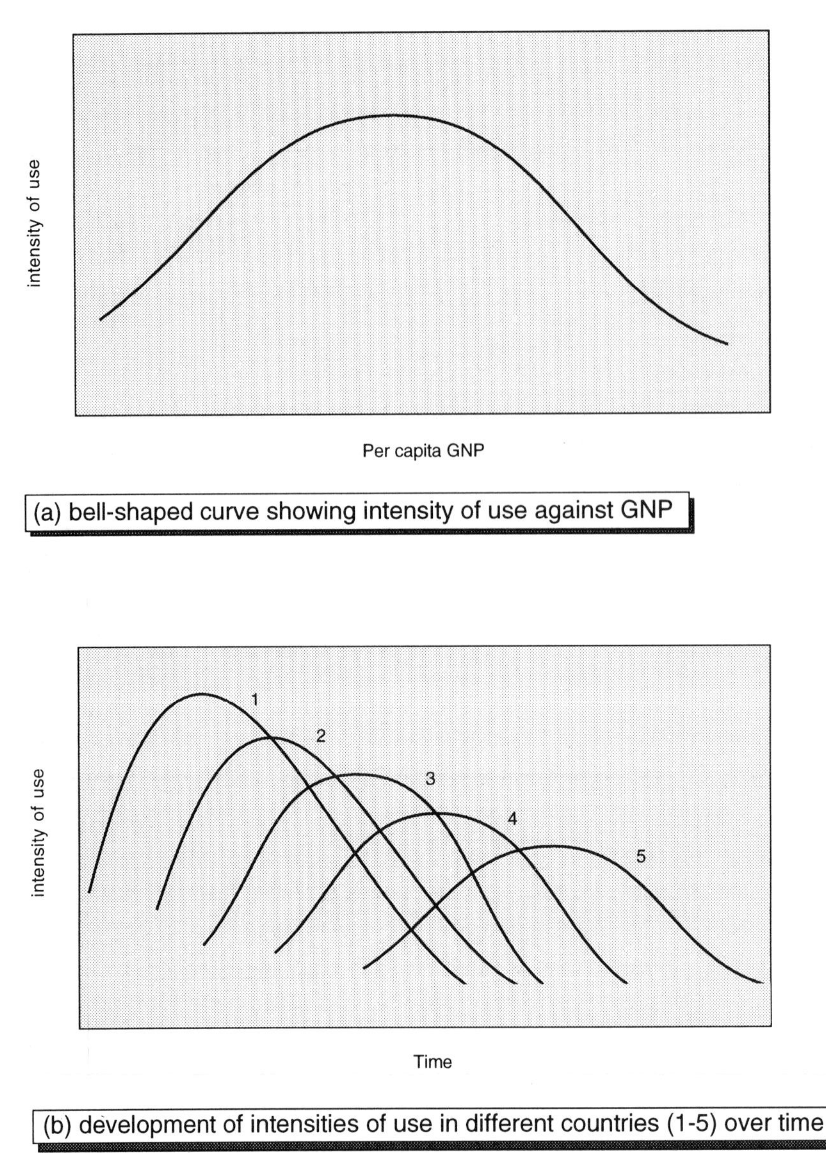

Figure 29 Aspects of the theory of 'natural' dematerialisation

material consumed per unit of GNP. He discovered that bell-shaped curves (Figure 29) govern the historical relationship between material intensity and per capita GNP for a range of different mineral resources in a broad cross-section of different countries.

The theory which emerged from this analysis – let us call it the theory of natural dematerialisation – has two main thrusts.[2] First (Figure 29a), it claims that material intensities peak during economic development and then begin to decline as GNP continues to grow. Second (Figure 29b), it argues that the peaks are lower, the later in time each country develops. This is believed to be because later developers are able to use technological innovations introduced by earlier developers to reduce their material intensities more rapidly than was possible for the earlier developers.

The theory of natural dematerialisation seems to be supported by some evidence of declining material intensities in the developed economies. But we need to treat the view that dematerialisation will happen of its own accord with caution for a number of reasons.

For a start, evidence of declining material intensities is restricted to certain specific materials. It is not immediately clear that we can extend the implication to the whole spectrum of material consumption, particularly as one of the driving trends of industrialisation has been a broadening of the material base, and increased complexity in the underlying chemistry. Iron and steel and concrete might exhibit reduced material intensities. But what about exotic metals, plastics, ceramics and petroleum-derived chemicals? Complex patterns of substitution make it difficult to generalise about overall material intensity from particular examples.

Next, we must take into account the tendency for industrialised nations to export their manufacturing activities to less developed nations and import finished products. This tendency will reduce the apparent consumption of raw materials in the developed nations. But it will do so only at the expense of increased raw material consumption – and increased environmental burdens – in the industrialising nations. It may be true that newer and cleaner technologies are available now which were not available in the earlier days of industrialisation. But we also need to remember that the conditions under which developing countries must industrialise today are vastly different and generally more hostile than those under which development took place for the Western nations. At any rate, irrespective of any decline in material intensities in the developed economies, the actual data do not yet warrant optimism about declining material intensities in the developing and newly industrialising world.

Finally, and perhaps most importantly, we need to question the relationship between material intensity and material throughput. The view that dematerialisation is happening automatically really only refers to the decline in **material intensities per unit of GNP**. Total material throughputs for the economy as a whole can only be determined after multiplying these intensities by the total economic activity (defined by GNP). Even if the amount of material consumed per dollar of GNP declines in certain places, the overall consumption of material may still rise if economic output is increasing.

THE NEED FOR POLICY INTERVENTION

These considerations highlight the urgency of engaging whole-heartedly in the strategy of dematerialisation. In particular (see Chapter 7), there is an urgent need to reorient the basis of profitability away from the throughput of material goods and towards the provision of services. The so-called 'service economies' which have evolved naturally in the industrialised nations over the last few decades are *not* service economies in this visionary sense.

It is one thing to articulate this vision. It is obviously quite another to implement it. We know from the previous chapter that the development of the vision has as much to do with commercial and institutional change as with technical innovation. And the lessons from relatively minor innovations in the energy supply industries suggest that this kind of change must be negotiated with care, commitment and collaboration. We cannot hope to achieve radical improvements in material efficiency without understanding the real economic and technical context faced by industrial operators. Extending product lives requires careful attention to the longer-term process of product and system design. Commercial innovation is – by its nature – bound to introduce institutional stumbling blocks and conflicts of interest which will require arbitration in a balanced and impartial forum.

For these reasons, it is almost unthinkable that the emergence of this new service economy could occur without government encouragement. It will also demand wide institutional involvement, committed industrial collaboration and active public participation. Perhaps the first step on this path will be to provide an effective forum in which these different actors can be brought together to negotiate change.

It is really beyond the scope of this book (and possibly of any book) to provide complete, unambiguous policy prescriptions for a new, and more sustainable, industrial economy. For a start, such an attempt would undoubtedly obscure the main aim of the book – to present a clear picture of what amounts to a paradigm shift in the industrial economy. More importantly, the process of policy implementation is – as I shall argue in more detail in the final chapter of the book – something which needs to be engaged in at a much broader level than could be done in a single-authored book. Nevertheless, there are a number of potential opportunities for procedural change, regulatory intervention and fiscal reform upon which some comment would be appropriate at this point.

AN OVERVIEW OF POLICY OPPORTUNITIES

In an earlier chapter I engaged briefly with two aspects of the policy framework through which governments are able to influence the behaviour of industries and consumers. One of these was **regulation.** The other was the use of **financial incentives** acting through market mechanisms of one kind and another. I shall have more to say about these two options later on. But these two kinds of policy measures do not exhaust the possible options through which the state can exert influence over individuals and firms.

For a start, governments inevitably form a **general policy framework** within which industries operate and consumers act. There are a number of ways in which this policy framework can influence environmental performance. Perhaps the single most important first step for governments in addressing the question of material throughput in the economy is to examine existing policies to determine what kind of impact is already being made by this policy framework. Transport policy is one of the areas which was highlighted in the previous chapter. The transport sector not only has considerable impact on the environment, it also offers significant opportunity for innovative system design. Much the same could be said of the energy sector, and the agricultural sector. Planning guidelines and local government policy are also important in reorienting communication networks and negotiating community development. A co-ordinated appraisal of these different policy areas could reveal numerous opportunities for preventive environmental improvement. It could also identify those areas where existing policies are operating against the aims of environmental protection.

The next step might be the formulation of a specific national **environmental policy plan** with clear long-term goals and interim operational targets. The Dutch National Environmental Policy Plan was probably the first example of such an initiative. Subsequently, there has been some movement towards the development of this kind of framework through government commitments to Agenda 21 – the long-term strategy for implementing sustainable development which was initiated at the United Nations Conference on Environment and Development at Rio in 1992. The development of these plans to include specific operational targets and clear strategies for implementation would undoubtedly further the environmental aims of dematerialisation.

The state also has a role to play in the provision of appropriate environmental and technological **information and training**. There is a wide range of audiences for whom education will be crucial. Governmental commitment to the provision of this education would have major long- and short-term impacts. Some of the specific measures which have been widely discussed are presented in Box 9.

Government **expenditure and procurement policies** can have significant direct and indirect effects on technology promotion and on environmental performance in the public sector. For instance, government might increase the share of recycled products that are purchased (where these products can be shown to be preferable from an environmental point of view), reduce its use of hazardous materials, install clean technologies in state-owned industries, improve energy efficiency in public sector buildings and so on. These initiatives have a multiplier effect, signalling government commitment, and reinforcing the credibility of technological improvements in environmental performance.

The strategies outlined in this book present quite specific demands for **research, development and demonstration** programmes. These demands cover a range of research areas. For instance, there are technological opportunities in almost every process industry and every industrial enterprise, large, medium or small. The development of more efficient and inherently cleaner process technologies requires a

BOX 9 INITIATIVES FOR IMPROVED INFORMATION AND TRAINING

- the promotion of general *environmental awareness* among consumers;
- *advertising campaigns* highlighting technical opportunities for reducing environmental burdens (e.g. through energy efficiency);
- the establishment of specific *advice centres* to provide technical support in the implementation of environmental improvements (e.g. energy efficiency);
- the provision of specific *environmental profiles* for consumer products and activities (e.g. ecolabelling)
- the establishment of *information clearing houses* for environmental information, technical assistance, and commercial opportunities;
- the *training* and co-ordination of specific environmental personnel (advisers, inspectors, information officers) with expertise in both preventive environmental management and process management;
- the development and re-orientation of *secondary and higher education* programmes to reflect an emphasis on preventive environmental management and appropriate environmental design;
- the development of *right-to-know legislation* on matters of environmental concern – e.g. the Toxics Release Inventory in the United States;
- the promotion (and possibly co-ordination) of *citizens' monitoring* initiatives to develop the active participation of members of the public in environmental policy.

commitment of resources in engineering design, which could offer significant environmental and economic dividends. In particular, of course, the reorientation of profitability away from products and towards services will demand completely new kinds of technology: for instance, those related to targeted dosing in agriculture or detergent recovery in cleaning operations.

But it is not just technical research which will be required. There is also a need for *methodological* research. For instance, techniques are needed which improve our ability to assess the environmental performance of different technologies, products and systems. The evolution of new commercial relationships will impose the need for new *economic* research. Issues of public acceptability and political feasibility will involve *sociological* disciplines. The emergence of new legal frameworks and organisational structures will demand *institutional* research. Governments have a role to play in providing and stimulating appropriate research and development in all of these areas.

THE ROLE OF REGULATION

Environmental **regulation** has traditionally played a major role in the policy framework which governments use to moderate industrial and consumer behaviour. Examples of environmental regulation include setting limits on emissions from industrial chimneys and pipelines; and banning the use of hazardous materials or their disposal in particular places. Because they are cast in terms of bans and limits, regulations are often seen as a rigid and sometimes inefficient way of achieving environmental protection. Generally speaking, regulation has gone hand in hand with the limited end-of-pipe approach to environmental protection.

But there are a number of other, more adventurous options open to governments using the regulatory framework. For instance, appliance and process efficiency standards can encourage technological improvement. Process standards can be related to **bench-marking** procedures, through which technology is continuously updated to improve environmental performance. Product standards could include requirements relating to modular design, component replacement and long life.

Flexibility can be built into these different kinds of regulation. The timescales over which they are to be achieved can be variable according to particular circumstances. And standards can deliberately be set – in advance – beyond current best practice, in order to encourage innovation.

There are other ways of encouraging and stimulating change without imposing inefficiencies on the system. Regulatory frameworks can be formulated which relate to operational procedures. In the United States, for example, the earliest formulation of **toxics use reduction** legislation required companies to audit their own facilities and produce their own targets for reduction. Another example from the US is provided by the Toxics Release Inventory (TRI). TRI legislation requires all companies over a certain size to provide annual data on emissions of a range of toxic and potentially toxic substances. Since the introduction of this legislation, many US companies have made increased commitments to reductions of TRI-listed emissions.

The formulation of product, process and material **liabilities** is another way of encouraging change in a flexible fashion through the regulatory framework. This has already been recognised, for instance by the EU's most recent environmental action plan. Recognising that 'it will provide a very clear economic incentive for management and control of risk, pollution and waste', the Community intends to establish an integrated approach to environmental liability, based in civil law. Traditionally, civil liability has been based on the need to prove fault, with a presumption in favour of the defendant. Increasingly, there is a demand for environmental liability to be based on a **strict liability** – independent of fault. Equally important, however, is the *scope* of liability legislation. Traditionally, environmental liability has applied mainly to damages caused as a result of industrial releases. There is now a rising demand for product liabilities. Usually these liabilities relate to defective products which either endanger health or do not perform as specified. Clearly, though, product liabilities could also relate to the environmental damage which products might cause, either during or after use. This kind of liability structure could be an extremely effective way of implementing the shift towards materials leasing. Users might lease material products from suppliers under certain conditions. But the suppliers would retain the ultimate responsibility for any materials supplied right through to the end of the product life.

In summary, then, regulations need not be seen as a set of draconian limitations on industrial activities, leading to inefficiency and loss of competitiveness. Rather there are a number of opportunities for creative intervention within a regulatory framework, through which governments can encourage and promote dematerialisation. In particular, the development of new concepts of liability could transform the structure of the market economy by shifting the basis of profitability towards the provision of services.

MARKET MECHANISMS

Traditionally, so-called market mechanisms (i.e. taxes and subsidies) have been seen as an alternative, perhaps even a competitor, to regulations. Debate over whether regulations or market-based instruments are the most appropriate means of ensuring environmental protection have been protracted and fierce. Regulations are usually seen as a more direct means of intervening to ensure environmental protection. Taxes and subsidies are often regarded as a more *flexible* way of encouraging improved environmental performance in a market economy. In reality, of course, it is inevitable that both regulations and financial incentives will be needed to implement and encourage the necessary changes. And in fact, the introduction of liability frameworks represents a blurring of the distinction between the two kinds of instruments. A regulatory framework imposes the liabilities. But they are enforced both through economic penalties for damage and through the payment of insurance premiums against future liability claims.

Whatever the relative advantages and disadvantages, there are certain reasons to suppose that the financial mechanism of **environmental taxation** is a strong candidate for consideration as an appropriate policy option. I first raised the issue of **environmental externalities** in Chapter 6. In particular, it was pointed out there that material prices are maintained at an uneconomically low level. Essentially, these prices are determined purely by the economic costs of extracting materials from the ground. In reality, this is a misleading basis for materials pricing because it does not include the external social and environmental costs associated with extraction, processing, and dissipation into the environment. Neither does this price basis make any compensation to the future for the depletion of scarce natural resources. So materials are generally priced below what might be considered economically optimal from society's point of view.

In the light of this, there is obviously a need to find ways of **internalising external costs**. That is, we need to try and ensure that the prices we pay for materials and material products include the costs of environmental damage and resource depletion which are associated with using and consuming those materials. There are a number of different ways in which this internalisation can be achieved. Both regulations and market mechanisms can be used. But one of the most obvious ways of internalising external costs is by using taxes.

In the ideal scenario,[3] the tax we apply to a particular material use or emission would reflect exactly the environmental cost associated with using or emitting that material. In practice, this ideal is difficult – if not impossible – to realise because of the problem of placing precise economic values on environmental functions (Plate 6).

Plate 6 How much for my penguin? – What is the value of a pristine environment?

Source: The Environmental Picture Library/© John Arnould

I have already commented on this difficulty in Chapter 6. In any case, the exact level of tax applied will, in reality, be influenced by a number of different factors, some of them political and institutional rather than purely scientific or economic. In this book, I am deliberately going to avoid making precise stipulations about these different factors. But I have also pointed out that failing to make *any* adjustments in price reduces the incentive to avoid polluting emissions. We could certainly argue, therefore, that our taxation system should be designed *to reflect the fact that* environmental damages result from material use and emission. And as the following section illustrates, there are significant advantages to be gained from an environmentally conscious reformulation of that system.

ECOLOGICAL TAX REFORM

Conventionally, governments have tended to raise many of their taxes on capital and on labour rather than on materials. Income taxes, local government taxes and national insurance or social security payments have provided the basis for state revenues in most Western nations. The difficulty with this strategy, from an economic point of view, is that it represents a significant distortion of the market. In particular, it means that labour is priced higher than the economic optimum. In the next chapter, we shall discuss in more detail how this distortion operates. We shall also see that rising unemployment is emerging as a critical structural problem within the economic system. So any strategy which reduces distortion of the labour market should be at a premium.

It seems as though this situation provides an almost perfect opportunity for change: increase the price of materials to reflect the social and environmental costs associated with them; decrease the price of labour by removing distorting taxes. **Ecological tax reform** – as it has come to be known[4] – provides one of the most promising policy options governments could hope to find, dealing simultaneously with two of the most pressing problems facing them.

Given that it seems to represent a win-win situation, it is perhaps surprising that it has not already been thought of, worked out and implemented in every democratic economy in the world. In practice, a few governments have made some preliminary attempts at it. In Sweden, for example, some taxes have been introduced on sulphur and nitrogen emissions from industry. But there are some difficulties. In the first place, governments can expect real opposition, both from materials producers and from materials consumers to rises in materials prices. Proposals to introduce an energy and carbon tax in the United States and in the EU were both effectively abandoned because of strong industrial resistance. Industrial lobbies argued that higher energy costs would seriously compromise economic growth and therefore threaten an increase in unemployment – exactly opposite to the intention of ecological tax reform!

Early economic models of the impact of carbon taxes seemed to bear this out. Taxes would have to be so high, claimed economists, that they would present industry with massive financial problems.[5] But these opponents and their early economic models neglected one crucial factor: in raising materials taxes, the governments would accumulate substantial revenues. These revenues could then be recycled in the economy in a variety of different ways. Later, more sophisticated economic models showed that **revenue recycling** was the critical issue in determining whether or not materials taxes represent an overall economic benefit or an economic burden. If materials tax revenues are recycled as reductions in taxes on labour, it is now agreed, then they can actually stimulate economic growth and increase employment.[6]

This, of course, is exactly what we would expect intuitively. But even with the endorsement of national economic benefits, governments may still face difficulties in implementing ecological tax reform. Even if the nation as a whole benefits from the shift, there may be short-term losers. One of the particular problems of recycling material taxes in the form of reduced income taxes is that it can have unpleasant effects on those who do not currently pay income tax. In the UK, for example, the most vocal source of opposition to proposals for a tax on domestic fuels has come from pensioners. In fact, the UK fuel tax proposal is not part of a considered package of tax reform measures. But even if it were, the same problem would arise. You cannot compensate non-wage-earners by reducing income-tax rates. Solutions exist, of course. A rise in pension payments would help. Tariff structures could be revised to reflect distributional concerns. Economic subsidies could be used to encourage energy efficiency and combat fuel poverty. But the point of the example is that *any* programme of tax reform requires considered, and carefully implemented government planning. Ecological tax reform is no exception.

Equally important to the long-term success of such a policy is a careful consideration of the impacts on materials suppliers. If the strategy is to be successful in reducing materials throughputs, then the output of the materials suppliers will certainly be affected. The fuel supply industries, the chemicals industries and the mining industries would all, at the very least, expect to suffer reduced physical output. In the long run, the same might be true for suppliers of other material goods. For instance, if durable consumer goods are made to last longer, the material throughput of suppliers of durable goods can only rise at the expense of reductions somewhere else in the market. The competitive producer of longer-lasting durables may be able to corner an increasing share of a limited market. But the material throughput of the sector as a whole must diminish – otherwise there will be no dematerialisation.

Pricing policies are clearly critical to the shift in profitability from materials supply to service provision. If materials are cheap and labour is expensive, then a producer will tend to be profligate with materials and thrifty with labour. Conversely, if the cost of materials increases and the cost of labour is reduced, there will be an economic incentive to increase employment and reduce material throughput. But ecological tax reform will not, on its own, achieve the desired transition. In fact, the likelihood is that it will not be politically possible to implement this kind of policy without other measures to encourage the shift from materials supply to the provision of services. There would simply be too much opposition from the many industrial lobbies whose profitability now rests on materials supply.

These considerations serve to illustrate the complex challenges which lie ahead if significant dematerialisation is to be achieved. On the other hand, the range of policy options from which governments can choose to influence commercial behaviour is broad, and the measures themselves are flexible. So our conclusion here must be that effective and comprehensive pricing of raw material inputs and environmental outputs is not beyond the wit of society or the policy of governments.

At the same time, if these strategies are to be successful in reducing overall environmental burdens we must still address the following critical question:

> Can improvements in material efficiency outpace (and continue to outpace) the increase in material throughput associated with economic growth?

LIMITS TO GROWTH

A World Bank paper has calculated that if developing countries are to reach the level of per capita income now enjoyed by Western countries there needs to be a 46-fold improvement in technological efficiency just to hold resource consumption and environmental emissions constant.[7] And another author argues that the industrialised countries themselves need to dematerialise their goods and services by a factor of 10, if they are to move on to a sustainable course.[8] Neither of these estimates allows for continuing economic growth in the industrial nations.[9]

Clearly, all the strategies outlined in earlier chapters of this book, and all the policies described in this chapter, will be of crucial importance if reductions in material intensity of this order are to be achieved. And it is possible that an intensive reorientation of the industrial economy along these lines could go some way towards meeting dematerialisation targets. But the final arbiter on the question of dematerialisation is once again thermodynamics.

Let us suppose, for the sake of argument, that it is possible to achieve a 50-fold reduction in material intensity over existing levels for all material throughputs, but that economic activity levels are allowed to grow indefinitely. The total material burden will fall for a while. But it will start to grow again once the level of economic activity exceeds fifty times its present level. In other words, if we insist on envisaging an economy which continues to grow indefinitely, then we must insist on a dematerialisation process which also continues indefinitely. And here we are led into something very close to an impossibility.

The second law of thermodynamics suggests that a certain minimum throughput of materials and energy is needed in order to maintain a complex organisational structure. The bigger the structure, the greater the requirement for maintenance energy. We can pursue certain kinds of efficiency improvements to minimise these requirements. But once we arrive at the limits of thermodynamic efficiency, increasing the level of activity implies increasing material dissipation.[10] In other words: **dematerialisation cannot go on indefinitely.** There are irreducible lower bounds to the material intensity of human activities. Once these bounds have been reached, an increase in activity levels signals an increase in material throughput.

In the light of this constraint, we need to revise our intellectual trajectory. It is vital of course to pursue the technological avenues for dematerialisation which have been outlined in the last few chapters of this book. But we must also be prepared to re-examine the underlying forces which have forged the industrial economy. In particular, the concept of economic growth has provided the springboard for the industrial economy, and continues to underpin it. And if we want to understand the full dimensions of the environmental problem, we must understand that concept.

9

GROWTH IN CRISIS
Untangling the logic of wealth

INTRODUCTION

Why do we want economic growth? If the material implications of continually expanding our economic base present so many environmental problems, why can we not simply stop the industrial train and get off? Surely, we in the developed nations have enough material wealth now? What is to be gained from pursuing economic growth any further?

There are really two main arguments for economic growth. First, the economic system that has been developed needs growth in order to ensure its own economic stability. Second, economic growth in the country is associated (in principle at least) with increased welfare in the population.

Each of these reasons would (if valid) be sufficient to ensure that the paradigm of economic growth were highly regarded – by economists, by politicians and by the nation at large. Together they have made that paradigm more or less unassailable. Despite a vast wealth of critical literature dating back to the industrial revolution itself,[1] the doctrine of economic growth is really one of the mainstays of the industrial economy. One need only pick up a newspaper, turn on the television or listen to the radio to witness the continuous preoccupation of industry, commerce and government with trends in the level of economic output. For these reasons, if for no other, it is important to understand the arguments for economic growth. By understanding them, perhaps we can get beyond them.

The second of these arguments is examined in more detail in the final chapter of the book. In this chapter, my primary concern is to address the first argument, the argument from stability. I have already remarked briefly on the relationship between employment and economic growth. Let us now pursue that relationship a little further.

GROWTH, DEMAND AND EMPLOYMENT

Generally speaking, the rise of capitalism in the industrial economy has been characterised by the pursuit of improved labour productivities; that is, by increasing the amount of output per worker. This was generally achieved by replacing human labour with machines, first in agriculture and in manufacture. Later, machines reduced workloads and increased outputs in the home, in industry, in transport, and in commerce. In the words of the Abbé le Blanc, an enthusiastic commentator on the industrialisation of eighteenth-century Britain, machines 'really multiply men by lessening their work'.

On the local level, one effect of this process was the wide-scale dislocation of the labour pool, on which I have already commented. It was this dislocation which prompted the civil unrest characterised most memorably by the Luddites. On the macro-scale, however, the surfeit of labour made available by the new machines was absorbed by new and expanding industries which were able to increase their profits for as long as the demand for their products was maintained. Ever since those early days, the goal of the industrialist has been to aim for continual improvements in labour productivity: more and more output per worker. From his or her point of view, the more output obtained per worker employed, the greater the income from sales, and the more profitable the enterprise.[2]

For as long as the nation as a whole can maintain full employment, this pursuit of improved labour productivity inevitably means that economic output increases. Conversely, of course, for as long as demand continues to grow, improved labour productivities do not lead to unemployment. But what happens in such a system when demand stagnates or reduces? In these circumstances, the systematic pursuit of improved labour productivities inevitably leads to reductions in the labour force. At the national level, this means unemployment. With more people unemployed, the spending power in the nation is reduced, serving to depress demand even further. Increased unemployment also means increased national spending – for instance on welfare payments. And this means the government must either borrow money or raise more in taxes. Borrowing money may increase long-term costs and reduce international confidence. Raising taxes reduces consumer spending further still, forcing the economy into a spiral of recession.

Dealing with this situation in the twentieth century has given rise to one of the fiercest and most divisive debates in the history of economics: that between Keynesian economists and the monetarists.[3] It is instructive to spend a few paragraphs analysing this argument because it provides a very clear illustration of the structural difficulties in which the industrial economy is still embedded.

The debate centres on the complex issue of unemployment.[4] Economic theory identifies a certain 'natural rate' of unemployment. This arises because of structural factors such as the continuous and inevitable change in the pattern of demand and production, and the existence of people whose physical or mental handicaps make them 'unemployable' in a job market geared towards specific forms of labour.[5] In addition, natural unemployment takes account of the impacts of trade union or organised labour pressure to enforce a minimum wage rate. But unemployment is also predicted to rise temporarily (in the eyes of the theory) when demand drops, until wages and prices have adjusted to a new equilibrium level.

According to the Keynesians, the appropriate response to 'demand deficiencies' is for the government to spend more money in order to stimulate new investment and generate new demand. This process is supposed to lift the economy into a new period of economic growth which will soak up the increase in the unemployed labour pool. The Keynesians gained almost complete ascendancy in economic thinking in the two decades following the Second World War. But they were drawn up short in the early 1970s by the biggest recession since the 1930s, characterised both by soaring inflation and by escalating unemployment. Suddenly, the Keynesian philosophy found itself incapable of answering the demands of the situation and came under attack from a new school of thought: monetarism.

The monetarists believed that government policy aimed at artificially inflating demand would disrupt the natural operation of the market and lead to unacceptable levels of inflation. Instead, the monetarists revived the same classical economic principles which Keynesianism had replaced thirty years or so previously. The centrepiece of their argument was the assertion that the market will itself solve the problem of unemployment and restore balance to the economy. As unemployment increased, they argued, this would drive down the wage rate, reducing the cost of labour to employers and allowing them to increase their production capacity by taking on more workers. This would simultaneously reduce unemployment, increase production and stimulate new demand. The appropriate role of government was just to reduce the response time of the market's natural tendency to rebalance itself at full employment.

The monetarists insisted, for instance, that governments should aim to reduce income taxes. Such taxes create a price differential between the level of wage an employer is prepared to offer and the level of wage a worker is prepared to accept. Equally, the monetarists argued, unemployment benefits should be reduced, because these reduce the incentive for workers to find work at the going wage rate. Additionally, monetarism perceived a need to reduce the power of trade unions, which might be able to negotiate a wage rate above the equilibrium, free-market wage.

The monetarists' philosophy was essentially one of non-intervention. By recalling the early doctrine of Adam Smith's 'invisible hand' (see Chapter 2), it provided the foundations for the 'free market' economics which has been pursued vigorously in most Western nations since the late 1970s, and whose effects will probably dominate the global economic and industrial climate well into the twenty-first century.

There are several instructive points to draw from this discussion. In the first place, neither the Keynesians nor the monetarists really contested that the key to maintaining full employment in the economy was to stimulate demand growth. Allowing demand to stagnate or fall was paramount, in their eyes, to allowing the economy to fall into a vicious cycle of unemployment, underinvestment, and recession – exactly the opposite of the virtuous circle represented by economic growth. As former British Prime Minister Edward Heath remarked: 'The alternative to expansion is not an England of quiet market towns linked only by trains puffing slowly and peacefully through green meadows. The alternative is slums, dangerous roads, old factories, cramped schools, and stunted lives.'

THE SCOURGE OF UNEMPLOYMENT

Neither the Keynesians nor the monetarists have really been able to understand or deal with the problem of rising unemployment levels. This problem is now moving to the top of the political agenda in the industrial nations. In 1993, the European Union (EU) published its White Paper on *Growth, Competitiveness and Employment*,[6] in which it set out targets for economic growth which would stabilise unemployment. An examination of the basis for that paper highlights the critical links between employment and growth in the market economy.

Obviously, if national labour productivity rises by 2 per cent per year (say) and the labour force remains constant, unemployment will rise unless output also rises by 2 per cent per year. In reality, the labour force[7] is still increasing slightly in Western nations, and these increases, together with the increased empirical labour productivities, mean that a growth rate of 3 per cent per year is now essential if unemployment in the EU is to be held constant. If growth fails to match this target, and productivity does not alter, unemployment will rise, the demand on the public purse will increase, and future economic output will be depressed by limitations on spending power. Unfortunately for the EU, the potential rate of growth – that is, the rate of growth which can be sustained without the economy 'overheating'[8] – is only about 2 per cent per year. Even in its own terms, therefore, the existing system is failing to deliver a sustainable economy.[9]

At the same time, accepting a rising rate of unemployment is tantamount to admitting that the free market is not doing its job properly. The monetarist response is to place much of the blame on organised labour which allows wage rates to be raised above the equilibrium wage in times of labour surplus. This 'overinflated' wage persuades industrialists to invest preferentially in capital rather than in labour, further increasing labour productivities and reducing the labour demand. But this is a dangerously regressive argument. The monetarist is saying that if we left things to Adam Smith's invisible hand, the problem would be solved because employers could continue to pursue their own profit by paying less money to their workers.[10]

Orthodox economic reasoning attempts to mitigate the moral inadequacy of this position by an argument which has become increasingly important to the environmental debate. This is the so-called 'trickle-down' philosophy, which I visited briefly in Chapter 5. This theory justifies the pursuit of economic growth on the grounds that the wealth created 'trickles down' from the richest investors to the poorest workers, and thereby acts to alleviate poverty at all levels. Although wage rates may fall temporarily during demand-deficient periods, the lower wage rates stimulate increased employment until a new equilibrium level is reached: fuller employment stimulates new economic growth, increased profits and wage rises.

Later in this chapter, I shall question the legitimacy of this position. But the problem of stability cannot be reduced to a question of social equity. Whether or not the monetarist policies of the 1980s are morally and socially acceptable, the fact of the matter is that neither monetarism nor Keynesianism has provided any kind of basis for dealing with increasing levels of unemployment. Each points to a single fundamental strategy: to increase demand growth to a sufficient level to offset improvements in labour productivity. Without economic growth the industrial economy heads quickly for a spiral of depression.

Are there ways of escaping this spiral? In principle, we might conceive of some ways of improving the situation. And ironically, it is from the direction of environmental policy that these possibilities emerge.

LABOUR AND MATERIALS: THE POTENTIAL FOR SUBSTITUTION

The EU White Paper observed candidly that Western economies 'over-consume Nature and under-consume people'. The substitution of labour by capital which has characterised the process of industrialisation 'has been accompanied by a continued increase in the use of energy and raw materials, leading to an over-exploitation of environmental resources', claimed the authors.

Intuitively, the obvious solution to this problem is to try and reverse the direction of substitution. Suppose we accept that the substitution of capital for labour has reduced much of the manual drudgery of our predecessors. Let us admit that the pursuit of increased labour productivities has powered the expansion of the industrial economy – with the help of an expanding demand base. Even so, can we not now say that enough is enough? If we have substituted a profligate use of materials for full employment, might not material efficiency help us to regain full employment? Could we not at least slow down the industrial machine's relentless progress towards the disenfranchisement of labour?

This, implicitly, is what the EU White Paper calls for. The authors argue specifically for an increase in the *employment content of growth*. And here is a strategy which finds many points of contact with preventive environmental management. In certain cases, the reduction in material throughput makes explicit demands on labour – for instance, to carry out improved maintenance and repair work on the factory floor. This suggests at least a possibility that firms could simultaneously increase their workforce, reduce their pollution and maintain their profitability.

What is true of the industrial process is equally true of the cycles of production and consumption which constitute the economic process. A major plank of the preventive environmental strategy (Chapter 4 again) is to improve materials utilisation through reuse, repair and reconditioning, remanufacture and recycling. This strategy – if effective – would make extensive reductions in the extraction and processing of raw materials. Reduced activity in the primary industrial sectors would mean the loss of some jobs in those sectors. An important part of government policy will be to find ways of compensating the losers and minimising the social impacts associated with economic change. But the improved utilisation of resources would tend to increase activity levels in manufacturing industry and in the service sector. In particular, there would be an increased need for skilled maintenance and repair work. A whole new sector of industry directed towards remanufacturing could also emerge.

The employment effect of this new manufacturing and service demand is reinforced by the relative materials and labour intensities of the different industrial sectors. Primary materials processing is energy- and pollution-intensive, but relatively low in labour intensity. By contrast, the manufacturing, service and repair sectors have low material requirements and high labour requirements. Once again, preventive environmental management offers the prospect of increasing labour intensity and reducing materials intensity at the same time.

Several important reservations should be made at this point. First, the economic viability of this reverse substitution of labour for materials is crucially dependent on the relative prices of labour and materials. This is why the strategy of ecological tax reform described in the previous chapter is important. A shift in the basis of taxation away from labour and towards the use and disposal of materials offers the possibility of double dividends: reduced pollution and increased employment. And even though such a policy is clearly interventionist, it intervenes in a manner that even the most committed free-market politician might agree with: the removal of distorting taxes and subsidies to improve the efficiency of the market.

If dematerialisation of the economy is to be successful, however, suppliers of material goods can only look forward to reduced material throughput. Nor is it obvious that they can offset reductions in physical output by increases in product prices. One of the main economic effects of dematerialisation will be to increase the supply of materials over demand, and – by the classic economic argument – to deflate material prices. Although governments could offset this deflationary effect by progressively increasing the level of materials taxes, this would only make matters worse for materials suppliers, who see none of the revenue from these taxes, and only experience a perpetually depressed demand for their products.[11] Demand saturation would then lead them into exactly the spiral of unemployment, underinvestment and recession so feared by modern governments. So the revision of profitability (see Chapter 7) envisaged under the new service economy again reveals its importance.

Finally, however, we return to the problem of limits. At best these strategies buy us time. Ultimately economic growth must lead us towards increased material throughput, in spite of efficiency improvements, and we are left face to face with the same problem. So let us revisit it from another perspective.

INTEREST AND SELF-INTEREST: THE WORKING OF THE PROFIT MOTIVE

The practice of paying and receiving interest on borrowed capital was condemned as usury by the Roman Catholic Church until 1830, and is still condemned by Islam today. But it is a crucial part of the modern capitalist economy. Suppose, for simplicity, that I operate a one-man industry. The basis for that industry will be a certain level of investment in what is called capital stock – machinery, buildings, and so on. In order to provide this capital stock, I will have to invest money in it. If I borrow £1,000 at a 10 per cent interest rate I will have to find an extra £100 to pay to my creditors in a year's time.

The only way of finding that extra £100 – other than reducing my own savings or salary – is to make sure that I make a profit from the venture. Essentially, this means that I must sell my output. And this leads to one of two things: either the total economic output in the country increases. Or else my output is sold only at the expense of a drop in someone else's output. So if my riches are not to make another poorer, then the profitability of my investment demands growth at the national level. If I fail to make a profit, I will be unable to pay off my capital, my creditors will demand compensation, and I will find myself out of business – unemployed.

What makes the situation worse is that I must continue to invest in my business. Thermodynamics insists that my capital stock cannot be maintained without the injection of new resources. So that next year, I will have to invest more money in my factory and increase my output further, if I am to make the interest payments. But increasing my output will in itself make new demands for capital stock, and this larger stock of capital will require a greater input of investment funds to maintain it. And so the situation continues year by year. If I do not grow, I do not survive. At the micro-economic level, the industrialist can only maintain his or her profitability by expanding output and by cutting costs. Historically speaking, industrial employers have vigorously pursued one particular means of reducing costs: improved labour productivity. These improvements lead to the macro-economic difficulties which I have already discussed.

Here then is the root of the problem. The structural stability of the industrial economy depends implicitly on increasing the economic output of the economy: economic growth is crucial to survival. What ties the two inextricably together is the practice of paying and receiving interest on capital.

But if this is the case, why could we not just do without interest? Why could we not re-endorse the age-old religious distrust of usury? In the long run, perhaps we could envisage that as a solution. In the short run, the problem is not so easily dismissed, for several reasons.

First, the most obvious practical measure in the direction of outlawing usury would be to reduce the interest rate. And ironically, this may well have a completely perverse effect. Reducing the interest rate would provide a two-way incentive for additional capital investment. For a start, capital would be cheaper for borrowers, who would therefore be tempted to borrow more, and invest more heavily in capital stock. Next, lenders would face reduced profits from the same outlay and would therefore be forced to expand their financial markets to include more borrowers if they wanted to survive. The overall impact of this reduction in interest rate might be to increase economic growth for a while. But it would do nothing to relieve the 'virtuous circle' of economic growth or the 'vicious cycle' of economic depression.

The simple response might be to advocate a zero rate of interest: to stipulate that no profit at all is derivable from lending out money. This would certainly cut the Gordian knot. Capital would be 'free' in a sense to borrowers. But no investor would lend under these conditions. At a stroke, the entire financial market would be rendered redundant. The systemic impacts would be potentially catastrophic. The structure of the global market-place is now tied firmly into the trade in capital. In many cases, it is the financial services sector of the developed economies which are growing fastest. And the development of the global political economy is really characterised by nothing more nor less than the extension of this same paradigm to every country in the globe.

The trouble is that, in the competitive, profit-driven market economy, growth operates in a Darwinian way. Only the fastest growing businesses and nations survive, and the alternative to growth in such an economic system is no longer the stable steady-state economies of pre-industrialisation, but economic collapse: a downward spiral of low output, underinvestment and unemployment.

In the late twentieth century, the economic forces governing the industrial economy are basically the same as those operating in the period of proto-industrialisation in Britain in the early eighteenth century. But the context in which those economic forces operate has profoundly changed. The pressures on industry to improve labour productivity remain unabated. But the reduced labour demand which these innovations foster is no longer cushioned by monopoly markets overseas. Unemployment rises. Social disintegration follows. Suddenly the future of the industrial economy is itself in question. The European Union – one of the most powerful trading blocs in the world – is talking of 'reversing the disastrous course which our society, bedevilled by unemployment, is taking'.[12] And the threat comes not so much from without as from within. The same driving forces which once created the conditions for industrial expansion are the ones which are now driving us towards economic and social collapse. The pursuit of economic profit forces us relentlessly onwards, and Adam Smith's invisible hand seems incapable of curbing its ill effects.

When orthodox economists allow us to question them on why we have developed and become so heavily implicated in such a system, the answer is remarkably straightforward. Profit is what makes us tick. It provides the engine for progress. It is the motor of development. Take away the profit motive and you take away the driving mechanism of human development, the *raison d'être* of civilisation.

In a sense, this reply takes us right back to where we started from at the beginning of the industrial revolution. This is the old, old, classical economic argument that self-interest is the motivator for human progress. This is the thesis which Smith first denounced (see Chapter 2) as 'holy [sic] pernicious' and later incorporated into the theory of the invisible hand. According to the classical economic argument, it is self-interest which prompts us to raise our collective standard of living, to harness new industrial resources, to expand our material horizons, and to improve our quality of life.

WEALTH vs WELFARE

This brings us back to the second of the two arguments for economic growth with which I started this chapter. According to this second argument, economic growth offers the prospect of continued improvements in human welfare. It is largely this equation of wealth with welfare which makes the pursuit of rising levels of GNP so attractive in political terms.

Clearly, however, increased wealth is not the same thing as improved welfare. We saw in Chapter 2 that the industrial revolution delivered somewhat mixed blessings in terms of human welfare, even in the early days. Some industrialists became very rich, very quickly, as a result of massive increases in output in certain industries. But many of their employees lived in conditions of appalling poverty, working in dangerous environments, and subject to economic forces over which they had absolutely no control.

Today, in advanced economies, we like to believe that we have overcome the 'growing pains' of industrialisation. But there are still a number of reasons why we should be suspicious of the equation of human welfare with measures of national wealth such as GNP.

In the first place, welfare is not solely determined by economic assets. Health, individual well-being, quality of life, environmental quality, individual and collective security, all make contributions to welfare which are not reflected in conventional measures of economic output like GNP.

Second, national income may be *spent* in a number of ways, some of which are purely 'defensive' rather than contributing additionally to welfare. An increasing proportion of income may be spent on cleaning up environmental damage resulting from the production of goods and services, or on treating illnesses arising from impaired environmental quality. For example, the costs of cleaning up oil spills (such as those from the Exxon Valdez and Braer tanker disasters) contributed to GNP. These kinds of defensive expenditure may be necessary to offset the adverse welfare effects of other expenditures. But it is then inappropriate to count both sets of expenditures as positive contributions to welfare.

Third, the notion of success in economic terms contains within it some notion of accountability for the future. It is not enough to have achieved a successful balance sheet for today, if this has been achieved by actions which render bankruptcy inevitable on the morrow. Recognising this need, the national accounts are sometimes adjusted to calculate the Net National Product (NNP), which measures the economic output net of capital depreciation.[13] But this adjustment is generally restricted to man-made capital: buildings, machinery, vehicles, etc. What about the depletion of what has been called **natural capital**:[14] mineral resources, clean air and water, fertile soil, diversity of species and so on? Figure 6 (in Chapter 2) showed how the industrial economy has become increasingly reliant on mineral resources. But no account is made of the depletion of these vital reserves of natural capital. Equally, no economic adjustment is made for the loss of environmental capital such as agricultural soils and pristine water supplies. As the Business Council for Sustainable Development describes it: 'The human species is living more off the planet's capital and less off the interest. This is bad business.'[15]

Ironically, many of these costs may not be felt by the present generation. Instead they are left as a legacy for our descendants. For example, the costs that will be associated with global warming from the fossil fuels emitted now may not be felt for several decades. Present industrial activity is living off the environmental quality of future generations. Accounting for long-term costs should also be an element of our attempt to measure overall welfare rather than present economic wealth.

Finally, aggregate measures of income may offer misleading indications of welfare if wealth is not evenly distributed throughout the population. Figure 30 presents an index of income distribution for the UK based on the so-called 'Gini coefficient'.[16] This number is a measure of the inequality of income between different income groups. The higher the Gini coefficient the less equal is the distribution of wealth.

The graph shows that the distribution of wealth remained more or less constant during the 1950s and 1960s, and even improved somewhat in the 1970s. But the index rose sharply through the 1980s,

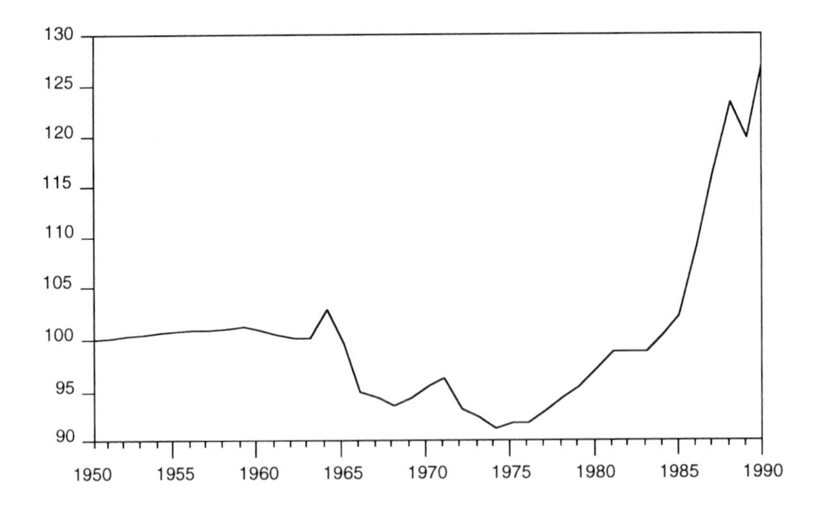

Figure 30 Inequality in the distribution of incomes in the UK: 1950–90 (The Gini coefficient is set at 100 in 1950 and subsequent years are indexed relative to the base year.)

indicating that the distribution of incomes was more sharply skewed towards the rich during that period. A dollar (or a pound or an ECU) in the pocket is worth more (in terms of welfare) to a poor family than to a rich one. So welfare cannot be said to be increasing at the same rate as income, if income is unevenly distributed. Furthermore, an uneven distribution of incomes may lead to social divisiveness, threatening personal security, and reducing rather than enhancing collective welfare.

But if GNP is an inadequate measure of welfare, for all these reasons, how exactly are we to judge whether or not economic growth is delivering improved welfare? One way of making that judgement[17] is to try and measure trends in welfare by adjusting GNP to account for a variety of economic costs which are not generally included in the analysis. The most recent attempts to do this are based on an Index of Sustainable Economic Welfare (ISEW) proposed by economist Herman Daly and theologian John Cobb in the United States.[18] This index attempts to measure welfare by adjusting an economic measure of consumer expenditure to account for a variety of environmental and social factors. Subsequently, the same methodology has been developed and applied to other countries. Figure 31 shows the indexed results of three of these preliminary attempts to measure sustainable economic welfare.

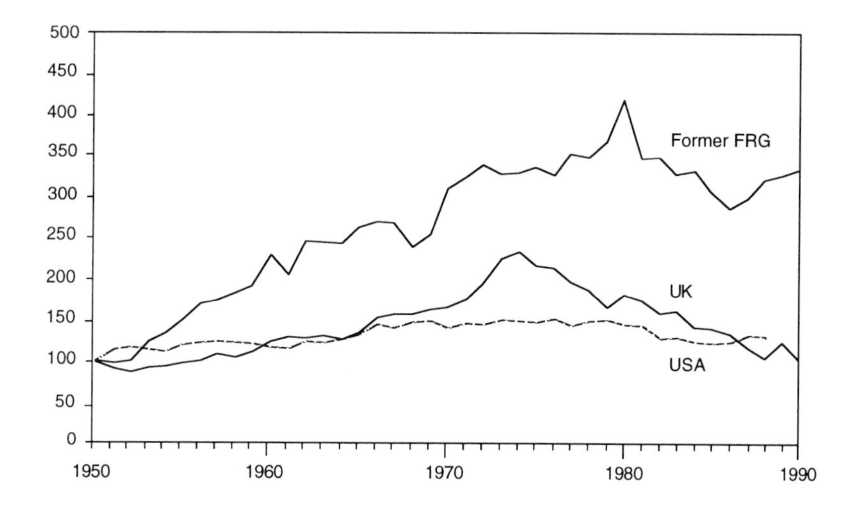

Figure 31 Trends in sustainable economic welfare in three industrialised countries: 1950–90 (The index is set at 100 in 1950 and subsequent years are indexed relative to the base year.)

Comparison of these results with Figure 8 (see Chapter 2) suggests that the trend in welfare departs significantly from the straightforward measure of economic performance over the four decades. In the UK, for example, per capita ISEW is barely greater by the end of the study period than it was at the beginning of the period, despite growth in GNP of over 200 per cent, and the index falls significantly through the 1980s, even though GNP continues to rise.

These results have to be regarded at best as illustrative. But the lesson from Figure 31 is nevertheless striking: welfare is not the same as economic wealth. Economic growth does not necessarily mean improved welfare.

DEVELOPMENT: A DILEMMA REVISITED

Despite all these observations, the pursuit of national economic growth – measured by growth in GNP – has been the common goal of national policy throughout the world, over many decades. The conventional development paradigm continues to argue that the creation of monetary wealth is the key to development. According to that viewpoint, it is this wealth which will provide us all with a higher standard of living. It is this wealth which will allow us to invest in new, more efficient technologies. It is this wealth – man-made capital – which can substitute wherever necessary for the losses of natural capital which are increasingly visible around us.

The same economic wisdom contends that the paradigm of global economic growth is the only way for poorer countries to develop. Arguing that 'rising tides . . . raise all boats',[19] economists see growth in the rich North as the best and probably the only way to alleviate Southern poverty. The 'trickle-down theory' of economic progress thus provides intellectual support for the translation of the consumer society into a global vision.

In reality, the conventional pursuit of economic wealth has failed to prevent increases in both relative and absolute poverty. The United Nations Environment Programme has estimated that the income disparity between the richest and the poorest 20 per cent of the world's population doubled between 1960 and 1990, while absolute per capita incomes in sub-Saharan Africa (for example) fell by almost 20 per cent during the 1980s.[20]

In fact, conventional economic development has failed to provide equitably for human needs even within the industrial nations. Problems of poverty in industrialised countries worsened during the 1970s and 1980s, according to the World Resources Institute, with the mean income of the lowest 20 per cent of the population in the United States falling by over 3 per cent between 1970 and 1990.[21] Figure 30 has shown the rapid changes in relative poverty experienced during the economic boom of the 1980s in Britain.

So what is the truth of the matter? Has economic growth made us worse off, or has it made us better off? Should we pursue it, or should we avoid it? How are we to make sense of the contradictory evidence we are faced with? On the one hand, economic development seems to have lifted us out of Boltzmann's desperate 'struggle for existence'. On the other hand, it has delivered untenable environmental impact, unsustainable resource depletion, and rising levels of systemic unemployment.[22]

There is no simple resolution to this dilemma. The complexity of the situation, and the extent to which we are embedded in it historically and socially, suggest that we will find no miracle cure, no simple palliative, no one-word answers to the problem. In a sense we have come full circle. This is almost exactly the dilemma which I outlined in Chapter 1 of this book; So maybe we should not be surprised that we have returned to it. What is perhaps remarkable is that the most advanced, most extensive and most accepted pattern of social organisation in human history should so self-evidently have failed to solve this dilemma for us. The industrial economy, as we have come to know it, is as much a part of the problem as it is a part of the solution.

This is not to say that the industrial economy itself is necessarily bankrupt. Even if the social and economic system we have come to accept as the norm proves to be unsustainable and inappropriate in the longer term, the reality is that for the time being at least we have to operate within it. In those circumstances, any strategies at all which can move us in the direction of a solution are at a premium.

Much of this book has been devoted to presenting options for reducing the material intensity of human activities. Many of those options are broadly technological and economic in nature. Is it possible that we could use the existing system to pursue these strategies? The answer is surely yes. Some institutional changes might be needed.

Profitability in such a revision would have to be based on something other than the volume sale of material goods. But the vision of a new service economy offers an exciting avenue for commercial development. Almost certainly, this process would buy us the time in which to devise a more appropriate framework for human development than the one which has presided over the economic, social and environmental crises we face today.

10

BEYOND MATERIAL CONCERNS

Regaining quality of life

INTRODUCTION

We are living in a material world. Life itself depends on fundamentally material processes. But our human lives are not, and have never been, solely and exclusively material concerns. This realisation has informed every religious doctrine in human history. But you do not need to be an advocate of any particular religion or creed to acknowledge it. All you need is an understanding of human behaviour which goes beyond the simplistic doctrine on which the industrial economy has built its own particular temple. It is now time to turn a critical attention on that underlying doctrine. First, though, let me briefly summarise where the investigation in this book now stands.

A RECAPITULATION

The dilemma that confronts human development is a fundamental one. The pursuit of economic growth has appeared to free modern society from the harsh struggle for existence which haunted earlier societies and which still characterises life in natural ecosystems. But the material profligacy of the industrial economy has brought with it unprecedented environmental degradation, and now threatens the long-term stability of our life-support systems.

How are we to deal with this dilemma? About the best that we have managed to achieve after 200 years or so of industrial development is the entrenchment of a well-worn conflict between economic rationale and environmental protection. Economic rationale demands continued growth; environmental protection demands constraint.

Capitalism insists on the pursuit of profit; environmentalism tries to stand in its way. The eventual result may turn out to be an already flawed system further deflected from its ideal path: not utopia but economic and environmental failure.

In these circumstances, there is a high premium on any strategy which offers a potential resolution to the stalemate. This is why the measures described in earlier chapters of this book are of such importance to the industrial economy. They present substantial opportunities for improved environmental performance which does not compromise economic competitiveness. And by doing so, the preventive approach extends a bridge across the chasm which traditionally separates economic rationality from environmental concern.

At the same time, the demands of the new approach – and in particular the substantial dematerialisation of the industrial economy which it implies – challenge the underlying logic of the prevailing economic system, a logic which has been with us since before the industrial revolution. That logic suggests two things. First, it presents the production and sale of material goods as the essential basis for commercial activity. Second, it insists on a critical structural role for economic growth. Taken together these two factors threaten to undermine the potential benefits of preventive environmental management.

We have seen that the first of these factors is ripe for reappraisal. The emphasis on material throughput is a contingent aspect of a particular historical development. It is not an immutable necessity. Chapter 7 has shown that there are clear prospects for reformulating economic profitability. The basis of this reformulation is the transition from profitability based on the sale of material goods to profitability based on the provision of a service. This transition demands that we revisit (see Chapter 8) the relative pricing of materials and labour, and revise the fiscal regimes which affect those prices. It also requires careful attention to regulatory frameworks and institutional structures. But these commercial innovations offer the prospect of double or triple dividends to the industrial economy: better economic performance, reduced unemployment, and improved environmental quality.

In the long run, we must also address the need for economic growth. As Chapter 8 explained, there are two powerful arguments for it. First, the structure of the existing economic system demands growth for its own stability. Second, wealth has become associated with welfare. A careful comparative analysis of national economic performance with welfare (Figure 31) reveals that, in reality, economic growth has failed to deliver consistent rises in welfare for a variety of reasons.

This is how the late Robert Kennedy once described it:

> The gross national product includes air pollution and advertising for cigarettes, and ambulances to clear our streets of carnage. It counts special locks for our doors, and jails for the people who break them. . . . And if the gross national product includes all this, there is much that it does not comprehend. It does not allow for the health of our families, the quality of their education, or the joy of their play. It is indifferent to the decency of our factories and the safety of our streets alike. It does not include the beauty of our poetry or the strength of our marriages, the intelligence of our public debate or the integrity of our public officials. . . . The gross national product measures neither our wit nor our courage, neither our wisdom nor our learning, neither our compassion, nor our devotion to our country. It measures everything, in short, except that which makes life worthwhile.

The conclusion itself is not very surprising. Most of us recognise that welfare is not the same as wealth. Most of us realise that the quality of our lives is not determined solely by our annual expenditure on stereos, televisions and portable phones. Not all of us would agree that some expenditures may actually detract from our lives. But almost all of us would generally rank material possessions below certain non-material factors: family life, friends, community. When the British Social Science Research Council asked 1,500 people what were the most important elements in determining their quality of life, 71 per cent of the replies they got had little or nothing to do with economic goods.[1]

At the same time, the pursuit of profit still grips the industrial economy in a cycle of material production and consumption. The stability of the economy rests on the profit motive. The pursuit of profit provides the philosophical foundation of capitalism. It furnished the spark which ignited the industrial revolution. It remains the dominant rationale for individual and corporate behaviour and for national development worldwide.

SCIENCE AND PSYCHE: A FAILED EXPERIMENT?

At this point in our investigation we have discovered what seems at first to be a startling phenomenon. The obstacles which now stand in the way of a reorientation of the industrial economy relate more to the psychology and sociology of human behaviour than to technical considerations or physical constraints. The orthodox economist might argue that these psychological facets of human behaviour are themselves akin to physical and technical constraints. Classical economics was founded with the very precise aim of mirroring for moral behaviour what deterministic mechanics had provided for the natural world. Adam Smith was a fervent admirer of Sir Isaac Newton. Nassau Senior, a follower of Smith, described the maximisation of wealth as a law of nature 'like gravity in physics', and in 1760 Helvetius wrote: 'as the physical world is ruled by the laws of movement, so the moral universe is ruled by the laws of interest.'[2]

In the eighteenth century, Newton's classical mechanics represented the pinnacle of human intellectual achievement, and its deterministic view of the universe reigned supreme. In the late twentieth century, this view of the world has been radically overturned by the theories of relativity, quantum mechanics and chaos. Even the physical world in which we must survive or perish is not really a mechanistic one. Uncertainty, relativity and irreversibility are its underlying characteristics. There is no simple determinism in the classical sense. What we learned from Newtonian physics has been valuable in many contexts, but it is not the whole truth.

So the first thing we can say about the economic 'quest' of the eighteenth-century classicists to provide a science of human behaviour is that it needs to be re-examined. We have learned a new understanding of the principles of classical mechanics. We need a new understanding of the principles of economics.

In fact, we could be considerably harsher about the entire economic enterprise. Even at the time, it was fundamentally flawed. The idea that you could create a science of transient human behaviour by analogy with the science of reversible physical motions would be laughable had it not turned out to be so disastrous. It is like supposing that you can develop a science of chess by studying the game of billiards. The rules of the game are fundamentally different. The analogy is empty.

Nevertheless, the economic construction which was built on the back of this attempt has endured for over two centuries. It has survived intact the revolutions which transformed physics, persisted through the explosive developments of modern psychology and sociology, and ignored the continuing censure of religious and spiritual leaders. The orthodox economist might argue[3] that here, if we needed it, was proof of the timelessness of the principle of economic self-interest. But this defence is a spurious one. Greed, avarice and vanity may be timeless enough. But the wholescale *institutionalisation* of the profit motive is not an immortal phenomenon. Its history can be traced. It belongs within the time-span of industrialisation in the modern world. It is part and parcel of a very specific development paradigm which has emerged only within the last two to three hundred years of a very long human history. It is virtually absent from the historical records of earlier societies, and is condemned by the surviving members of indigenous peoples to this day.

In any case, there is something more important to say about history than this. At the end of the twentieth century we face a point of critical decision. Confronted by economic and social instability on the one side and escalating environmental degradation on the other, our paradigm of development is itself in crisis. The roots of that development are all inextricably entwined around the principles of economics. Unless we are prepared to revisit and reassess those principles, we are unlikely to solve our environmental and social problems.

SELF-INTEREST

The fundamental claim of classical economics was that self-interest is more beneficial to the general advancement of society than good intentions are. As Adam Smith himself put it in *The Wealth of Nations*: 'It is not from the benevolence of the butcher, the brewer, or the baker, that we expect our dinner, but from their regard to their own self-love.' Another early economist, David Ricardo, specifically drew the parallel between self-interest and the profit motive. But it was John Stuart Mill who was responsible for the idea of the self-interested **rational economic man** as the basis for economic development.

Mill was heavily influenced by the utilitarian philosophy of Jeremy Bentham, a close friend of his father. **Utilitarianism** proposed that people do the things which are most likely to bring them the maximum pleasure and avoid the things which are most likely to bring them pain. According to the utilitarians, pleasures did not differ between themselves in kind, but only in strength or intensity. This became an important element of the new science of human behaviour, because it insisted that happiness (or utility) could be measured in a single currency. For the economists, that currency was a monetary one.

Mill himself was profoundly troubled by the implications of the emerging viewpoint. Shortly before he died in 1873, he made a crucial distinction which ought logically – had it come a little earlier – to have sunk the entire enterprise. 'Those only are happy,' he declared in his *Autobiography*, 'who have their minds fixed on some object other than their own happiness; on the happiness of others, on the improvement of mankind, even on some act or pursuit, followed not as a means but as itself an ideal end.' The implication is clear. The pursuit of the profit motive may deliver wealth. But the pursuit of happiness is unlikely to deliver happiness. Wealth and happiness are different in kind. And the pursuit of profit cannot by its nature be expected to deliver happiness.

These reservations remained on the periphery of the new economics. Instead, economic self-interest emerged as the guiding principle of human development. Its pedigree stretched back in time through Bentham's utilitarianism at least as far as Greek **hedonism**. And the extension forwards in time of the same philosophical and psychological doctrine reaches well into the twentieth century: for example, through Skinner's **behaviourism** and Freud's **pleasure principle**.

Perhaps in the long run, we will have to reject the utilitarian view of human nature if we want to move convincingly beyond the conflicts into which classical economics has led us. On the other hand, we can hardly deny that the doctrine of self-interest is firmly embedded in our cultural traditions. Whatever the deficiencies of classical economics, we cannot entirely dismiss the doctrine of self-interest. As the economist will argue: people *do* seem to be motivated, at least in part, by self-interest. They *do* care whether their wages rise or fall.

They *are* concerned about improving their material welfare. Material possessions provide security for themselves and their families. Change and instability frighten and confuse them. Improving their status in life promises stability, provides a sense of progress, and satisfies a desire to leave something behind them for their children. What is more, continues our enlightened utilitarian, a part of that self-interest is a laudable concern for the well-being of one's dependants and descendants. We might even have to admit that self-interest (in this broader sense) underlies our concern about the environment.

And if self-interest is so patently present in the world, should we not attempt to organise society in harmony, rather than in conflict, with it? Even supposing that the narrow view of self-interest incorporated into classical economics is misguided, might there not be a different, and more acceptable, view of our underlying psychology on which to base society? Is self-interest truly served by a narrow pursuit of economic wealth? Is there a version of self-interest flexible enough to accommodate a new vision of human development?

Suddenly, these questions assume increasing importance as we unravel the dynamics of the industrial economy. Answering them comprehensively is really beyond the realm of this book. But we know enough now to hazard a guess at some of the answers a more thorough investigation might uncover.

WELFARE AND HUMAN NEEDS

Let us return to a very basic view of what society is. Clearly it has something to do with ensuring collective and individual welfare. The economy is a system for providing the services (see Chapters 4 and 7) which deliver that welfare. This begs the question of what welfare is. If we spurn the conventional economic assumption that consumption provides a proxy for welfare, then we have to face the difficult question of what exactly we do mean when we speak of it.

There are a number of possible perspectives on this. One of them is the monetarised index shown in Figure 31. But if we are rejecting the equation of wealth with welfare, we might also regard the *monetarisation* of welfare with some suspicion. We could supplant, or perhaps supplement, this kind of measure with certain **development indicators** to reflect different aspects of welfare. Such indicators could be provided, for instance, by measures of life expectancy, infant mortality, access to clean water supplies, literacy and so on. These are all useful ways of measuring our success or failure in delivering specific services. But we have still not answered the question of what we mean by welfare.

A constructive way of looking at the problem is through the satisfaction of needs. Human needs have been the subject of continuing investigation since the work of Abraham Maslow in the 1950s and 1960s. There is now an emerging consensus that human needs are both specific and identifiable.[4] This same consensus also suggests that it is possible to characterise common, underlying needs which can be attributed universally to all human beings. This assumption is not essential for what I want to argue, but it is certainly useful in limiting the amount of ground we cover.

Using this characterisation, the provision of welfare can be represented as the process of satisfying our underlying human needs. Poverty, by contrast, is perceived as the failure to satisfy needs. One of the implications of this observation is that there are as many different kinds of poverty as there are different kinds of needs. This insight allows us a completely new perspective on the question of development. Under the conventional notion, development aims to eliminate financial poverty. Defining poverty in monetary terms also defines economic growth as the only solution. But if we accept the idea that there are many different kinds of poverty and that the aim of development is to provide for many different kinds of needs, then we are in a different position altogether. Economic growth can no longer be expected *per se* to eliminate poverty. Nor should we be surprised if economic growth has failed to provide unequivocally for human welfare.

More importantly, material consumption cannot be expected to alleviate poverty either. Nor should we be surprised if material consumption has failed to provide for human welfare. The reason is simple. As we shall see below, not all human needs are material ones.

In order to take these ideas any further, we clearly need to look in more detail at human needs. What are needs? What kinds of differences characterise different needs? How may these different needs be satisfied? What happens if they are not satisfied? Can we substitute the satisfaction of one need for the satisfaction of another? Before addressing these questions, let me just comment on the way in which orthodox economics handles the question of needs.

Conventionally speaking, economics translates all needs as 'wants' or consumer preferences and then makes the grand assumption that these preferences are expressed by the monetary values of goods and services traded in the market.[5] This is a very powerful equation. It allows the economist to declare that the market economy is by definition satisfying human needs, because people buy and sell only what they want (i.e. need) to buy and sell. If welfare is defined in terms of the satisfaction of needs, then clearly the market economy is automatically providing for human welfare because human beings have expressed those needs (as preferences) through the market. Equally, economic growth then becomes the route to welfare precisely because the common currency of the market is a monetary one. The monetary economy allows consumers access to all the needs (equals wants) that are traded on the market – always provided they can afford to pay for them.

There are a number of reasons to dismiss this equation.[6] First, it does not help us to understand why the market has failed to deliver either environmental quality or freedom from material poverty. Second, there are certain kinds of entities which are not traded, and probably not tradeable on the market, and it would be shortsighted to forgo the possibility that some of these entities are the object of human need. Here are some examples: peace, tranquillity, friendship, freedom, creativity, beauty, environmental quality. Next, the English language has provided us with a distinction between needs and wants and it seems ungracious to deny it. And finally, the elision is also likely to suppress our understanding, because there *is* clearly a distinction between needs and wants. A person may not want to eat; but as a biological organism he or she needs to eat. If that person does not eat, eventually they will die.[7] Equally, a person may want to commit murder, but to suggest that that want is equivalent to a 'need' to commit murder is to destroy all pretence at moral or social imperatives.

Let us accept, then, that wants are not needs, in spite of the ferocious attempts of marketing agencies, advertisers, profiteers and racketeers to persuade us that they are. Let us take instead a more considered approach to the question of welfare, based on the idea that human

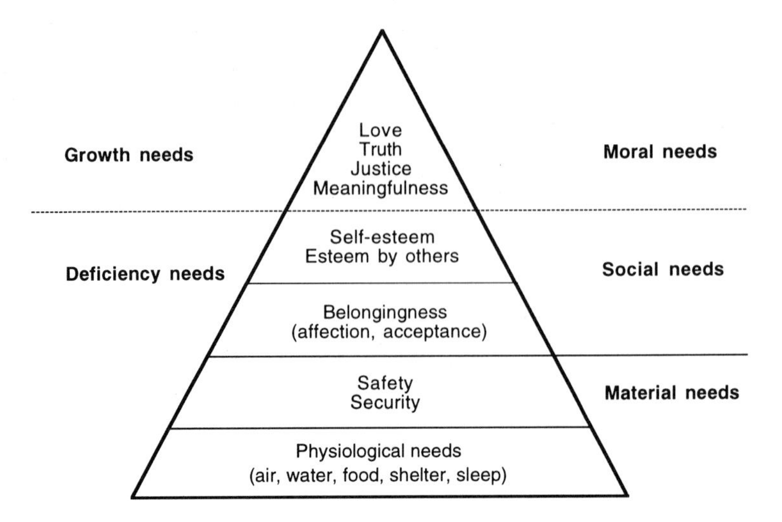

Figure 32 Maslow's hierarchy of human needs

beings possess specific identifiable underlying needs. What are these underlying needs?

The best-known characterisation is due to Abraham Maslow, whose early work[8] defined a hierarchy of needs stretching from basic physiological needs to what Maslow called moral needs at the top of the hierarchy (Figure 32). Maslow's idea was that the higher needs – which he characterised as **human growth needs** – remained latent until the lower needs – which he characterised as **deficiency needs** or maintenance needs – had been satisfied.[9]

The idea of a hierarchy of needs possesses a long historical pedigree. For instance, Plato declared in *The Republic* that 'the first and chief of our needs is the provision of food for the existence of life. . . . The second is housing and the third is raiment.' And there is a sense in which this hierarchical aspect follows common sense. If human beings do not have adequate nutrition, they die; and our success or failure in satisfying any other needs is then irrelevant.

But there are also troublesome aspects to this hierarchy, which have attracted a degree of criticism of Maslow's work. For example, the hierarchy has been interpreted as an argument for economic growth. Since higher needs are supposed to be latent until the deficiency needs are met, personal growth appears to be dependent on reaching a

		Being	Doing	Having	Interacting
Material needs	Subsistence				
	Protection				
Non-material needs	Affection				
	Understanding				
	Participation				
	Identity				
	Idleness				
	Creativity				
	Freedom				

Figure 33 Human-scale development: matrix of fundamental human needs

sufficient level of personal wealth. This interpretation almost certainly runs counter to the spirit of Maslow's theory. In fact, in his later work he revised the strict hierarchy of Figure 32 to place growth needs and deficiency needs side by side.[10] This dual hierarchy was designed to reflect what he saw as a duality in our human nature.

Later investigations also try to avoid the hierarchical interpretation. The particular characterisation I want to borrow here comes from a project on 'human-scale development' inspired by the Chilean economist Manfred Max Neef and co-ordinated by the Dag Hammarskjöld Centre in Sweden and the Development Alternatives Centre (CEPAUR) in Chile.[11] The project identified nine fundamental human needs. These nine needs are envisaged as occurring in four 'existential' categories: being, having, doing and interacting. The emerging matrix (Figure 33) of thirty-six distinguishable categories of fundamental needs is in stark contrast to the single, commensurate measure of utility which underlies classical economics.

There are clear parallels between this characterisation and the work of Maslow. For instance, the subsistence needs correspond to the basic physiological needs in Maslow's model. Security and protection are closely related. Many of Maslow's 'social needs' are mirrored in the Max Neef framework. Although some of the growth needs in Maslow's hierarchy appear to be absent in Figure 33, Max Neef has suggested a tenth need for transcendence which might bear many of these characteristics.

NEEDS AND SATISFIERS

One of the most important elements of this emerging body of work is the crucial distinction between needs and **satisfiers**. Each culture attempts to satisfy collective and individual needs in its own way. The underlying needs may be universal. But the satisfiers vary from culture to culture and from time to time. In our own culture we have chosen a very specific development path characterised by the system of production and consumption in the industrial economy. The fact that this same development path is emulated on an increasingly global scale does not deny its cultural relativity. It only suggests that we should be even more careful to examine the appropriateness of the underlying paradigm.

The industrial economy arose from within an economic system designed primarily to provide for the material needs of subsistence and protection. Satisfying these vital needs involved the supply of certain material commodities: foodstuffs, clothing, construction materials. Industrialisation was based on the extension of this network of trade in vital material commodities to a wide range of other material products. Mass production of material goods had to be matched by a mass market for industrial products. The success of industrialisation was its ability to expand those markets, and create new ones for its material products. But what is the relationship between these new, mass-produced material products and fundamental human needs? Because the remaining needs are essentially non-material ones, it is no longer clear that the appropriate extension from the provision of physiological needs is the provision of more material products. Affection, creativity, idleness[12] and participation are radically different kinds of 'commodities' from those which the market economy was designed to deliver.

The 'human-scale development' project (amongst others) made a critical distinction between the provision of economic goods and the satisfaction of needs.[13] In the conventional orthodoxy, economic goods are supposed to provide what people want, as expressed by the transactions of the market. But what is the relationship between economic goods and human needs? Because we are rejecting the idea of a common currency of utility, expressed through financial transactions in the market, we have to accept the possibility of a much looser relationship between economic goods and the satisfaction of needs.

Some economic goods may actually satisfy wholly or partly a particular fundamental human need. **Singular satisfiers** both aim at and achieve the satisfaction of a single need without inhibiting the satisfaction of other needs. **Synergic satisfiers** simultaneously satisfy or contribute to the satisfaction of more than one need. Breast-feeding is a possible example of a synergic satisfier. It simultaneously satisfies the child's need for subsistence and its needs for protection, affection and identity.

Not all economic goods are satisfiers in the singular or synergistic sense. Some may only bring a false and temporary sense of satisfaction to a particular need. Examples of these **pseudo-satisfiers** may be junk food (subsistence), prostitution (affection) and fashion (identity and participation). Other kinds of economic goods (**inhibiting satisfiers**) satisfy one need only by impairing the satisfaction of other needs.

There is worse to come, however. The work of the Dag Hammarskjöld project suggested that there may be some economic commodities which fail to satisfy any fundamental needs at all and in fact only destroy the possibility of satisfying them. An often quoted example of such a **violator** is nuclear weapons. Supposedly set in place to satisfy the need for protection, the arms race has impaired the satisfaction of subsistence, affection, participation and freedom. Ultimately, it even threatens the satisfaction of the protection need.[14]

Clearly, we are now in an area of investigation which is highly subjective in its nature. What satisfies one person (or set of people) may not satisfy another, even though the underlying need is the same. The satisfaction of needs is strongly determined by culture, class, social expectation, and individual choice. For these reasons, this new approach to development insists that the satisfaction of needs is something which requires negotiation in an appropriate interpersonal framework. It highlights the *local* identification of needs, and the *local* determination of appropriate satisfaction. This contrasts starkly, of course, with the centralisation of production and government which has accompanied industrial development. But it is reflected in some emerging political priorities: for instance, in the principle of subsidiarity on which the European Union rests; and in the 'Local Agenda 21' initiatives to implement sustainable development at the local level which emerged from the Rio Conference on Environment and Development in 1992.

MATERIAL IMPLICATIONS

Perhaps the most important point of all about these investigations is the division between *material* needs and *non-material* needs. This division is explicit in Maslow's model. It is also inherent in the Max Neef framework. Subsistence and protection needs are inherently material needs. Satisfying them implies the need for material activity – primarily the provision of food, water, clothes and homes. Of course, there are more and less materially intensive ways of producing food, supplying water, providing clothes and building homes.[15] Nevertheless, the existence of these vital physiological needs implies a certain minimum requirement for material activity. As we would expect, the human economy remains a fundamentally material concern: material inputs and outputs, obeying fundamental physical laws.

The striking feature of the remaining needs is that they are *not* inherently material in nature. The satisfaction of these needs is not entirely independent of material activities. But there may be a wide range of options for satisfying them. Some attempts will be highly intensive in term of material consumption. Others will be less intensive. Because the needs themselves are non-material, the satisfaction of them does not obviously dictate a particular level of material activity. Rather, the material throughput associated with satisfying non-material needs depends on cultural choice about appropriate satisfiers.

This is an absolutely vital point to have reached in our investigation. The whole book has been concerned with the potential for a dematerialisation of human activities. If we take the existing system of production and consumption for granted, we limit our intervention to technological improvements in process efficiency, and materials recycling. By introducing the idea of institutional changes, we open up the possibility of new commercial relations in the service economy. By questioning the fundamental principles of the system we reveal an immense new field of opportunities.

In the industrial economy we have tried almost exclusively to satisfy these non-material needs in material ways. Material possessions are used in the industrial economy to reflect status, to provide identity, to prove participation, to create a sense of belonging, to improve creativity, to strive for freedom and even to emulate affection. In some cases, material goods attain a limited measure of success in meeting these non-material concerns. But often they act as inhibitors, pseudo-satisfiers and violators.

In Chapter 7 I discussed the provision of appropriate transportation services. I suggested there that there were more and less materially intensive ways of supplying passenger-kilometres. But I also indicated that the supply of passenger-kilometres was not the ultimate objective of transportation. In terms of the discussion in this chapter we could now argue that transportation is in fact just one way of attempting to meet a number of underlying human needs. Transport services get us to and from work. Work provides us with subsistence, participation and identity. Mobility also gives us access to our friends and relatives – supplying a need for affection and a sense of identity and belonging. Automobiles themselves are more than just vehicles of transportation. They are also symbols of freedom and identity.

Some of the needs provided by transportation are material ones. Others are non-material. But the present system is very intensive of materials use. It may satisfy some of our needs. But often it operates as a pseudosatisfier. The loss of environmental space associated with an expanding roads network inhibits the satisfaction of leisure needs, and violates our need for quality in the environment.[16] At times – as congestion on the roads becomes more and more problematic – the system even fails to satisfy the identity and freedom needs promised by the iconoclastic motor car. There is no freedom and little personal identity attached to 'gridlock'!

There are ways of revising the transportation sector. Some of these involve technological alternatives. Others involve reassessing the demand for passenger-kilometres, and reconceiving our communities to reduce the need for travel. But our attempts to devise new systems will not succeed unless we address all the needs which the existing system promises to supply. There is no point in restricting private transport without providing accessible and affordable public transport. Equally, there is no point in discouraging mobility without encouraging better communities and improved participation. Perhaps most importantly, there is no point damning the car until we understand its status as a satisfier of needs. Freedom, identity and creativity needs are as important as participation and subsistence. If they are to be successful, our new communications networks must find – or perhaps simply allow for – new ways of meeting all of these underlying needs.

The important point is this: many of the needs which we are currently attempting to satisfy through materially intensive systems are non-material needs. But the relationship between material goods supplied by this system and non-material needs is no longer clear. In fact, it never was clear. Mass production of economic goods was a misconceived enterprise when viewed from the point of view of fundamental needs. And the social implications of this misconception are as alarming as the environmental ones. Environmental degradation and resource depletion may just about be excusable, if they are fundamentally necessary for the provision of human welfare. But if material profligacy destroys the environment only to serve us with violators and false satisfiers, then it quite clearly represents an unforgivably foolish development path.

BEYOND MATERIAL CONCERNS

Here at last we have arrived at what appears to be a core misconception in the heart of the industrial economy. By basing itself on a narrow concept of self-interest translated in terms of economic wealth and the possession of material goods, modern society has accepted a kind of poisoned chalice. Offering sanctity of choice, fulfilment of our desires, and the greater good of our fellow human beings, it has delivered environmental destruction, economic instability and new, alarming kinds of poverty: poverty of identity, poverty of community,[17] and poverty of spirit.[18] As Lewis Herber remarked, we have reached 'a degree of anonymity, social atomization and spiritual isolation that is virtually unprecedented in human history'.[19]

What characterises the development of the industrial economy, and perpetuates the myth of the success of the consumer society, is the assumption that all human needs can be satisfied through material goods. Accepting the existence of fundamentally non-material needs shatters that myth. But the prognosis for the future is brightened.

Traditionally, the struggle for sustainable development has been perceived as a conflict between environmental quality and human welfare. This is the dilemma with which I started out in Chapter 1, and which we have revisited continually throughout this book. But the concept of human welfare which has brought us to this dilemma has been too narrow: our vision of self-interest has been a misleading one. The system we have constructed to satisfy our needs has led us astray: an economy profligate of material resources, an environment degraded by material emissions, and lives overburdened by material concerns. From within the darkness of this position appears the light of a new conception: a society which addresses itself anew to the complex question of satisfying needs; an economy which does not degrade the environment in which it must survive.

In this book we have only glimpsed at the possibility of this emerging vision. Moving towards it will be no simple task. Our culture has chosen a very specific development path. Its attempts to meet our underlying needs have potentially devastating material implications. How many of these attempts are fundamentally flawed? How much of the material throughput associated with them represents pseudo-satisfiers, inhibitors and violators? How many of these needs could be met in less materially intensive fashion? How many of them could be met in entirely non-material ways?

We do not yet know the answers to these questions because we have not been accustomed to addressing them. But we do not have to accept these features as immutable consequences of the physical world. They are choices. These choices are now embedded in powerful institutions and complex social frameworks. But they are nevertheless choices. We cannot change the laws of thermodynamics. But we can influence and exercise cultural choice. And this strategy may well provide us with the most significant and extensive opportunities for dematerialisation that we could hope to find.

Most importantly, we do not need to perceive new choices as a threat to our economic or physical survival. On the contrary, if we have the courage and the wit to devise a society in which non-material needs are once again recognised, and appropriately addressed, we can only expect to improve the quality of our lives.

NOTES

1 LIVING IN A MATERIAL WORLD

1 The distinction between renewable and non-renewable resources is not always straightforward. For instance, many of the processes which give rise to deposits of non-renewable resources continue to occur naturally, even if the speed at which resources are deposited is very slow by comparison with the speed at which resources are extracted. Equally, the so-called renewable resources may often be harvested by society at a rate faster than they are replenished through natural processes, sometimes leading to deforestation and the loss of diversity in the natural environment.

2 This sector is also sometimes called the commercial sector or the service sector. In a later chapter of this book, the implications of these different names for the tertiary sector will become clearer. In fact, they form the basis for a reconsideration of the historical division between economic sectors.

3 1 megajoule (MJ) is equal to 1 million joules. A joule is a unit of energy measurement.

4 These figures are taken from I. Bousted and G. Hancock, *Handbook of Industrial Energy Analysis*, Ellis Horwood, Chichester, 1979.

5 In fact this law is only approximately true. Because of the equivalence of mass and energy (as expressed by Einstein's famous equation $E=mc^2$) there are situations in which mass is transformed into energy and vice versa. In these circumstances the conservation law applies to the total balance of energy/mass equivalents.

6 During nuclear transformations, some individual elements change from one type to another, so that this kind of conservation law does not hold. Instead it is replaced by different conservation laws. But the complexity of these nuclear processes is beyond the scope of this book. In global terms, these kinds of processes still represent a relatively minor contribution to the material activities of the industrial economy.

7 The interested reader will find a full discussion of the implications of

the second law in P. Coveney and R. Highfield's (1991) book, *The Arrow of Time*, Flamingo, London. A more detailed discussion of the importance of the second law for the interaction between economy and environment is given in T. Jackson (ed.), *Clean Production Strategies: developing preventive environmental management in the industrial economy*, Lewis Publishers, Boca Raton, FL, 1993.

8 A. Eddington, 1953, *The Nature of the Physical World*, Cambridge University Press, New York.

9 Georgescu-Roegen was really the first to highlight the importance of the second law of thermodynamics for economic processes. His book (*The Entropy Law and the Economic Process*, Harvard University Press, Cambridge, MA, 1971) is fascinating, if complex, reading.

10 This interpretation is due to the physicist Boltzmann, who devoted much of his life to the formulation of statistical thermodynamics. The Boltzmann equation which expresses it was engraved on his tombstone in Vienna.

11 A system is closed – but not isolated – when it can interchange energy with its environment but cannot interchange materials.

12 This view has been developed within thermodynamics by the Nobel laureate Ilya Prigogine and his co-workers (see I. Prigogine and I. Stengers, *Order out of Chaos*, Heinemann, London, 1984) and also within ecology – most notably by E. Schrodinger (*What is Life? The physical aspects of the living cell*, Cambridge University Press, Cambridge, 1945), A. Lotka (*Elements of Physical Biology*, Williams & Wilkins, Baltimore, 1925, reprinted as *Elements of Mathematical Biology*, Dover, New York, 1956) and more recently by H.T. Odum (*Systems Ecology*, John Wiley, New York, 1982).

13 A system is open when it can interchange both energy and matter with its environment.

14 The word 'organic' has two important – and unfortunately entirely unrelated – uses in environmental literature. The sense in which it is used here derives from the discipline of organic chemistry which is chemistry based on the element carbon. The second use of the term 'organic' refers to practices and procedures which are holistic in nature, and attempt to utilise natural ecological processes as much as possible.

2 MATERIAL TRANSITIONS

1 This kind of knowledge base and conduct code characterises those cultures previously operating on a subsistence basis and remnant now only within so-called indigenous tribes and populations.

2 See, for example, Debeir *et al.*, *In the Servitude of Power: energy and civilisation through the ages*, Zed Books, London, 1991, Chapter 5.

3 See, for example, the discussions in C. More, *The Industrial Age*, Longman, London, 1989, Chapter 9.

4 Like the industrial revolution, the term 'capitalism' is also subject to different interpretations, but this one will do for our purposes. There is little disagreement on the principal elements of the capitalist system: the pursuit of profit, the development of a market economy, the private ownership of property, and the accumulation of investment capital (see More, *Industrial Age*, p. 416, for example).

5 Adam Smith – sometimes called the father of economic science – published his seminal work *The Wealth of Nations* in 1776. But his *Theory of Moral Sentiments*, in which he proposed the doctrine of the 'invisible hand', was published seventeen years earlier in 1759. And Smith drew considerable inspiration from seventeenth-century thought. In particular, he was greatly influenced by the mechanistic thinking of Isaac Newton, and strove to emulate for the moral universe what Newton had provided for the physical universe, a set of deterministic laws which governed and predicted social behaviour.

6 Mill's *Principles of Political Economy* was published in 1848; but the doctrine of utilitarianism on which it was based belonged to his father's teacher and friend Jeremy Bentham.

7 Of course, it was also present as the antagonist in Marx's celebrated attack on the capitalist system in 1860.

8 Debeir *et al.*, *Servitude of Power*, Chapter 5.

9 Figures cited in this paragraph are taken from Debeir *et al.*, *Servitude of Power*; from E. Hobsbawm, *Industry and Empire*, Penguin Books, London, 1990; and from figures quoted in W. Humphrey and J. Stanislaw, 'Economic Growth and Energy Consumption in the UK, 1700-1975', *Energy Policy*, vol. 7, pp. 29–42, March 1979.

10 This includes **renewable energy** consumption – about 20 per cent of the total – as well as nuclear resources and fossil fuel resources which amount to just under 10 billion tonnes of coal equivalent.

11 The term 'energy intensity' usually has a quite specific meaning, namely the energy consumed per unit of Gross National Product. In this context, I am using the term slightly more loosely to refer to the per capita energy requirements in different societies.

12 The exceptions to this rule are perhaps those domestic animals and agricultural livestock which enjoy some of the fringe benefits of anthropogenic energy surplus.

13 See Table 12.2 in More, *Industrial Age*.

14 The World Resource Institute's *World Resources 1994-1995* (Oxford University Press, New York and Oxford, 1995) shows a slight decline in ore consumption between 1987 and 1992. But a significant increase over previous years.

15 This figure is taken from the author's own analysis based on data supplied

by the World Bank's Industrial Pollution Projection System. It is essentially a very rough estimate, but provides some kind of lower-bound estimation of global toxic emissions.

16 The halogens are the chemicals chlorine, bromine, fluorine and iodine. Apart from the threat which these gases present to the ozone layer, halogenated organic compounds (i.e. compounds containing both a halogen and carbon) offer particular environmental threats in all media.

17 E. Goldsmith and N. Hildyard's *The Earth Report 3: an A–Z guide to environmental issues* (Mitchell Beazley, London, 1992) is a good starting place for the unfamiliar reader.

18 See Figure 13 in Hobsbawm, *Industry and Empire*.

19 Bernard de Mandeville, *The Fable of the Bees* (1714), vol. 1, Oxford University Press, Oxford, 1966.

20 Cited in R. Douthwaite, *The Growth Illusion*, Green Books, Bideford, 1992, p. 36.

21 It has been argued by some historians that this process was only possible because of the massive demand created by overseas markets in which indigenous producers were overwhelmed and outclassed. In a sense, then, development occurred in the industrialised nations only at the expense of poverty in the colonies. The colonies were also crucially important, of course, for the provision of raw materials which fed the industrial process. Cheap material resources and cornered markets were the building blocks of economic growth in Europe. See R. Douthwaite's excellent book *The Growth Illusion* for a detailed critique of the economic and social history of industrialisation.

3 FAREWELL TO LOVE CANAL

1 Details of the story of Love Canal have been taken from a number of different sources, including Goldsmith and Hildyard, *Earth Report 3* (1992), and a report in the *New York Times* from 22 October 1990.

2 For a fuller discussion of the costs of cleaning up contaminated sites in the US, see Chapter 5 in Jackson, *Clean Production Strategies*.

3 This viewpoint is most clearly underlined by the consideration of economic and ecological systems as *non-equilibrium* thermodynamic systems (see Chapter 1). Such systems tend to exhibit certain kinds of behaviour which are inherently unpredictable because very small changes in initial conditions can lead to very large changes in the state of the system. This feature of climatic systems is one of the reasons it is so difficult to provide accurate weather forecasts, for example.

4 The term 'organic' is being used here in the sense of relating to carbon chemistry. The particular organic form of mercury which gave rise to the Minamata Bay incident was methyl mercury.

5 In 1984, the US National Academy of Science found that there was no toxicity information at all on 77 per cent of chemicals in commerce and only 'minimal' information on the remaining 23 per cent (see National Academy of Science, *Toxicity Testing: strategies to determine needs and priorities*, National Academy Press, Washington, DC, 1984).

6 This sensitivity can be defined in terms of a concept known as the **critical load**: the environmental load (for a particular pollutant) below which no harmful environmental impacts occur. There are clear similarities between the concept of critical load and the concept of assimilative capacity. Both assume that there is a level of emissions into the environment which is essentially safe.

The historical difference between the use of these concepts is an interesting one, however. Generally speaking, the assimilative capacity concept has been used as a **permissive principle**, justifying the release of certain contaminants into previously pristine environments. The critical loads concept has mostly been used, however, as the scientific justification for a call for wide-scale reductions in emissions of acid pollutants. This is because existing environmental loads are generally far above the so-called critical loads.

There is not really any paradox here. It is simply that when actual loads are above the calculated critical load (or exceeding the assimilative capacity) the thrust of both principles is prohibitive of further emissions. But when actual loads lie below the calculated critical load (or within the assimilative capacity) the thrust of both principles is permissive. The danger in both cases arises from the difficulties of accurately determining what those loads and capacities are.

7 This option has been adopted for instance by National Power, the UK's largest power utility, as the 'best practicable environmental option' for its planned 'orimulsion' (a cheap bitumen-based fuel) power station at Pembroke (*ENDS Report*, 236, September 1994).

8 The flue gas desulphurisation at Drax power station in North Yorkshire is expected to consume 600,000 tonnes of limestone a year (*ENDS Report* 228, p. 7).

9 Although gypsum is used in the building trade, the quantities generated are liable to create an early glut on the market. The 4,000 MW Drax power station will produce around a million tonnes a year, for instance. In addition, power station gypsum tends to suffer from contamination by heavy metals, leading to doubts about its suitability for the construction trade (*ENDS Report* 228, p. 7).

10 For a full discussion of this incident, and in particular the complex question of immunocompetence in fish, see V. Dethlefsen, T. Jackson and P. Taylor, 'The Precautionary Principle – towards anticipatory environmental management', Chapter 3 in Jackson, *Clean Production Strategies*.

11 See E.P. Odum, *Basic Ecology*, Saunders College Publishing, Philadelphia, PA, 1983, p. 47.

12 Early descriptions of these approaches can be found in a number of places. A particularly useful overview is provided by the US Office of Technology Assessment's *Serious Reduction of Hazardous Wastes*, 1986. Other sources include: M. Campbell and W. Glenn, *Profit from Pollution Prevention*, Pollution Probe, Toronto, 1982; J. Hirschhorn and K. Oldenburg, *Prosperity without Pollution: the prevention strategy for industry and consumers*, Van Nostrand Reinhold, New York, 1991; INFORM, *Cutting Chemical Wastes: what 29 organic chemical plants are doing to reduce their hazardous wastes*, INFORM, New York, 1985; and the 1990 proceedings of a US EPA International Conference on Pollution Prevention, *The Environmental Challenge of the 1990s, Clean Technology and Clean Products*, US Environmental Protection Agency, Washington, DC.

13 See, for example, S. Maltezou, A. Metry and W. Irwin, *Industrial Risk Management and Clean Technology*, Verlag Orac, Vienna, 1990.

14 Cleaner Production was the title used by a UNEP industry and environment programme to describe 'a conceptual and procedural approach to production that demands that all phases of the life-cycle of a product or process should be addressed with the objective of prevention or minimisation of short- and long-term risks to human health and the environment' (see L. Baas, H. Hofman, D. Huisingh, J. Huisingh, P. Koppert, F. Neumann, *Protection of the North Sea: time for clean production*, Erasmus Centre for Environmental Studies, Erasmus University, Rotterdam, 1990).

4 A STITCH IN TIME

1 It could certainly be argued that some kinds of goods are provided *for their own sake*. Even in this case, it is possible to construe this as a provision of services. I shall return to this point at a later stage in the book.

2 The term 'organic' is being used here in the second of the two ways identified in Chapter 1, namely to refer to procedures and practices which are holistic in nature.

3 This diagram is adapted from procedures outlined in the UNEP (1991) Technical Report Series no. 7, *Audit and Reduction Manual for Industrial Emissions and Wastes*, UNEP, Paris; and from the Alaska Health Project guide: *Profiting from Waste Reduction in Your Small Business*, Alaska Health Project, Anchorage, AL, 1988. Both of these publications are invaluable for small companies setting out to identify pollution prevention opportunities.

4 Note, however, the discussion relating to this process change in Chapter 6 below.

5 In the diagram it is assumed that the hazardous material in question goes to the aqueous waste stream. But it might as well have gone to air emissions or to the solid waste stream.
6 It has also reduced a number of other environmental burdens associated with raw material extraction and processing.
7 For an overview of energy efficiency trends in the different sectors of use, see L. Schipper and S. Meyers, *Energy Efficiency and Human Activity: past trends and future prospects*, Cambridge University Press, Cambridge, 1992.
8 These four strategies are examined in more detail in a number of other places. See, for example, Chapter 14 in Jackson, *Clean Production Strategies*; see also the (1982) OECD report on *Durability and Product Life Extension* and W. Stahel's (1986) paper 'Product Life as a Variable', in *Science and Public Policy*, vol. 13, no. 4, pp. 185–93.
9 For further discussion of the cascade concept see W. Stahel and T. Jackson, 'Optimal Utilisation and Durability', Chapter 14 in Jackson, *Clean Production Strategies*. See also R. Clift and A. Longley, 'An Introduction to Clean Technology', Chapter 6 in R. Kirkwood and A. Longley (eds), *Clean Technology and the Environment*, Blackie A&P, London, 1995.
10 This process is sometimes called away-grading or down-grading.
11 In the absence of appropriate avenues for recycling materials, a further possibility is the recovery of energy from the calorific content of the materials.
12 For more detailed discussions on the concept of material intensity per unit of service (MIPS) the reader is referred to a special edition (vol. 2, no. 8) of the *Fresenius Environmental Bulletin*, 1993.

5 EASY VIRTUES

1 Reported in Campbell and Glenn, *Profit from Pollution Prevention.*
2 Reported in the *Wall Street Journal*, 11 June 1991.
3 Reported in INFORM, 1985, *Cutting Chemical Wastes*, by D. Sarokin, W. Muir, C. Miller and S. Sperber, INFORM, New York.
4 Reported in INFORM, 1992, *Environmental Dividends: cutting more chemical waste*, by M. Dorfman, W. Muir, and C. Miller, INFORM, New York.
5 *ENDS Report* 228, p. 7.
6 Reported in *Clean Production Worldwide*, UNEP, Paris, 1993.
7 Reported in Hirschhorn and Oldenburg, *Prosperity without Pollution.*
8 I am grateful to Professor Roland Clift for suggesting this schematic representation.
9 Apart from the costs of the equipment and the costs of the limestone to operate the system, flue gas desulphurisation reduces the conversion

efficiency of the power station, implying increased costs for the additional coal required.

10 See T. Jackson (ed.), *Renewable Energy – prospects for implementation*, Butterworth-Heinemann, Oxford/Stockholm Environment Institute, 1993.

11 See WCED, *Our Common Future*, Oxford University Press, Oxford, 1987.

6 PERSISTENT VICES

1 Some of this literature is summarised in T. Jackson, *Efficiency without Tears – 'no-regrets' energy policy to combat climate change*, Friends of the Earth, London, 1992. The book by Schipper and Meyers, *Energy Efficiency*, also discusses some of the impediments to energy efficiency.

2 For example the United Nations Environment Programme's Industry and Environment Programme Activities Centre in Paris operates the computerised database ICPIC as an international clearing house for information on pollution prevention in industry.

3 This summary of full-cost accounting – sometimes also called *total-cost accounting* – is adapted from the Alaska Health Project report and from Chapter 9 of Jackson, *Clean Production Strategies*, and references therein.

4 For example, see M. Jacobs, *The Green Economy*, Pluto Press, London, 1991, and M. Sagoff, *The Economy of the Earth*, Cambridge University Press, New York, 1989.

5 Others are cadmium, arsenic, lead, chromium and antimony.

6 The most vociferous lobby for a chlorine phase-out is Greenpeace International, who have campaigned for a complete ban on chlorine production. See, for instance, J. Thornton's *The Product is the Poison: the case for a chlorine phase-out*, Greenpeace, Washington, DC. See also 'Chlorine under Pressure', p. 4 in *The Chemical Engineer*, 9 March 1995.

7 'Pathogen' is a term applied to disease-carrying or disease-causing micro-organisms such as those which are present in raw sewage.

8 One of the reasons for the historical preference of the chloralkali process over lime causticisation is that the feedstock for the former (brine) is considerably more plentiful than the feedstocks for the latter (lime and soda ash).

9 Another, more worrying manifestation of the problem of lost revenues for bulk materials producers is the opening up of overseas markets. As environmental regulations tighten in the industrial world, there is an increased pressure on bulk materials producers to find new markets elsewhere. All too often this leads to the transfer of materials and material products which have been banned in industrial nations to overseas markets, where environmental regulations are less stringent.

7 BACK TO THE FUTURE

1 C.P. Kindleberger, *Economic Growth in France and Britain*, 1964, p. 158 cited in Hobsbawm, *Industry and Empire*, p. 41.
2 Hobsbawm describes the battle for power between these two classes in his *Industry and Empire*. Before the industrial revolution power and social status resided with the merchants – giving rise to the description of Britain as a 'nation of shopkeepers'. This basis shifted considerably during and immediately after industrialisation.
3 See Hobsbawm, *Industry and Empire*, p. 31.
4 It was the enclosure movement which gradually began to give the manufacturers more of a constituency in government, because it gave the politically decisive group of landowners direct and widespread interests in the mines.
5 The concept of a 'new service economy' has been developed most extensively by O. Giarini and W.R. Stahel in *The Limits to Certainty*, Kluwer Academic Publishers, Dordrecht, 1989.
6 This is most obvious by considering the utilitarian roots of economic theory (Chapter 2). Money in economics served only as a proxy for a concept of 'utility'. Maximising expected utility was the way in which the philosophers translated the idea of 'bringing the greatest happiness to the greatest number'.
7 See Figure 5.4 in Douthwaite, *Growth Illusion*.
8 And I would certainly caution against uncritical acceptance of the claim that tourism is a legitimate route to sustainable development for developing economies.
9 The economic evidence for this is now overwhelming. See, for instance, the papers in T. Johansson, B. Bodlund and R. Williams (eds), *Electricity – efficient end-use and new generation technologies and their planning implications*, Lund University Press, Lund, 1989.
10 See Jackson, *Efficiency without Tears*.
11 This diagram is taken from an analysis by C. Cicchetti and W. Hogan (*Including Unbundled Demand Side Options in Electric Utility Bidding Programs*, Energy and Environment Policy Center Report no. E-88-07, Kennedy School of Government, Harvard University, Cambridge, MA, 1988), which is discussed in more detail in Jackson, *Efficiency without Tears*.
12 More details of this kind of commercial innovation can be obtained from the Product-Life Institute in Geneva.
13 These implications and the advantages of them have been pointed out in a number of places. See, for example, W.R. Stahel, 'Product Life as a Variable', *Science and Public Policy*, vol. 13, no. 4, p. 185, 1986; O. Giarini and W.R. Stahel, *The Limits to Certainty: facing risks in the new service economy*, Kluwer Academic Publishers, Dordrecht/Boston/London

(2nd edition) 1991; Stahel and Jackson, 'Optimal Utilisation and Durability', in T. Jackson (ed), *Clean Production Strategies*.

14 See M. Börlin and W.R. Stahel, 'Die Strategie der Dauerhäftigkeit', Bankverein vol. 32, Schweiz Bankverein, Basel, 1987.

15 See P. Nieuwenhuis and P. Wells (eds), *Motor Vehicles in the Environment*, John Wiley, Chichester, 1994.

16 For a more detailed discussion of these aspects of 'car culture' see, for instance, P. Freund and G. Martin, *The Ecology of the Automobile*, Black Rose Books, New York, 1994.

17 This is one of the roles played by travel agents for package holidays but it is virtually unknown for domestic travel.

18 The data on which this figure is based come from a study on *Ecolabelling Criteria for Washing Machines*, carried out by the UK Ecolabelling Board, London, 1992.

19 This example is taken from Clift and Longley, 'Introduction to Clean Technology', in Kirkwood and Longley, *Clean Technology and the Environment*, where it is discussed in more detail.

20 See J. Corbett, K. Wright and A. Baillie, *The Biochemical Mode of Action of Pesticides*, Academic Press, London, 1984, p. 343.

21 See P. Johnson, 'Agricultural and Pharmaceutical Chemicals', Chapter 7 in Kirkwood and Longley, *Clean Technology and the Environment*, p. 225.

22 This example is also discussed in more detail in Clift and Longley, 'Introduction to Clean Technology', in Kirkwood and Longley, *Clean Technology and the Environment*.

8 NEGOTIATING CHANGE

1 See Debeir *et al.*, *Servitude of Power*.

2 See, for instance, O. Bernardini and R. Gallini, 'Dematerialisation: long-term trends in the intensity of use of materials and energy', *Futures*, vol. 25, pp. 431–48, May 1993.

3 At least according to the conventional neo-classical economic model. For a fuller description of this model see, for example, D. Pearce, A. Markandya and E. Barbier, *Blueprint for a Green Economy*, Earthscan, London, 1989.

4 For a more detailed exposition see E. von Wiezsacker's and J. Jesinghaus's book *Ecological Tax Reform*, Zed Books, London, 1992.

5 Given the huge potential for *profitable* reductions in energy consumption from improved energy efficiency, this argument is slightly confusing. But I have already discussed (in Chapter 6) some of the reasons for this paradox.

6 Most of this analysis is to be found in quite technical papers on economic modelling of the impacts of carbon taxes. For the interested reader, the

following are examples: F. Laroui and J. Velthuijsen, 'The Economic Consequences of an Energy Tax in Europe: an application with HERMES', SEO Foundation for Economic Research, Amsterdam, 1992; J. Proops, M. Faber and G. Wagenhals, *Reducing* CO_2 *Emissions: a comparative input–output study for Germany and the UK*, Springer-Verlag, Berlin, 1993.

7 Robert Goodland and Herman Daly, 'Why Northern Income Growth Is Not the Solution to Southern Poverty', Environment Department Divisional Working Paper no. 1993-43, World Bank, Washington, DC, 1993.

8 See F. Schmidt-Bleek, 'MIPS – a universal ecological measure?' *Fresenius Environmental Bulletin*, vol. 2, pp. 306–11, 1993.

9 For estimates which do, see the paper 'Environmental Sustainability and the Growth of GDP' by P. Ekins and M. Jacobs in A. Glyn and Z. Bhaskar (eds), *The North, the South and the Environment*, Earthscan, London, 1995.

10 The only way we could avoid this conclusion would be by postulating an economy in which economic output grows without any increase in physical activity levels. Although in theory it might be possible to devise an economic system which fulfilled this strange condition, it is clear that the present system does not oblige. Economic output as measured by GNP is – as Herman Daly points out in his seminal *Steady State Economics* (Earthscan, London, 2nd edn, 1992) – a value index of a *physical* flow. 'In calculating GNP, efforts are made to correct for changes in price levels, in relative prices, and in product mix, so as to measure only real change in physical quantities produced.'

9 GROWTH IN CRISIS

1 In 1857, for instance, John Stuart Mill wrote:

> I cannot . . . regard the stationary state of capital and wealth with the unaffected aversion so generally manifested towards it by political economists of the old school. I am inclined to believe that it would be, on the whole, a very considerable improvement on our present condition.

The most vociferous contemporary proponent of the steady-state economy is former World Bank economist Herman Daly. See also Douthwaite's trenchant critique in *Growth Illusion*.

2 This drive for 'reduced labour content' can also be driven by the desire to reduce product prices. Generally speaking, however, reduced product prices are an attempt to capture a wider market, and thereby expand sales output.

3 The Keynesians were followers of John Maynard Keynes, whose views

on demand-determined output and employment were first published in *The General Theory of Employment, Interest, and Money* in 1936. The monetarists were generally followers of Milton Friedman and the 'Chicago school' of economists which emerged in the 1970s.

4 Useful discussion of the underlying theory can be found in many economics textbooks, for instance: D. Begg, S. Fischer and R. Dornbusch, *Economics*, McGraw-Hill, London, 1991.

5 This kind of employment is usually called frictional unemployment.

6 *Growth, Competitiveness and Employment – the challenges and ways forward into the 21st century*, White Paper, Commission of the European Communities, Luxembourg, 1993.

7 The labour force includes potentially employable people who are actually unemployed for various reasons.

8 'Overheating' in the economy means that high levels of inflation replace rises in real output.

9 For a more detailed discussion of this question in relation to European Union economic policy see David Fleming's paper 'Towards the Low-Output Economy: the future that the Delors White Paper tries not to face', *European Environment*, vol. 4, part 2, April 1994, pp. 11–16.

10 In fairness, I should say that the monetarist response would also point to the need to reduce distorting taxes and benefits in the economy. But the reality is that income taxes cannot really be held responsible for the continuing rise in unemployment for the simple reason that they themselves have remained constant or fallen. Equally, unemployment benefits have fallen in real terms: see Table 27.4 in Begg *et al.*, *Economics*.

11 These remarks suggest that some questions of international equity are raised by the dematerialisation of Western economies. Many of the raw materials required for the manufacture of products consumed in the West come from developing countries. A decline in the consumption of material products in the West would reduce the demand for these resources, and thereby reduce the export potential of raw materials producers.

12 *Growth, Competitiveness and Employment,* White Paper, Commission of the European Communities.

13 Nevertheless, GNP is more generally used as an indicator of economic and political success, and as an index of welfare, than is the NNP.

14 The reader unfamiliar with this term will find plenty of discussion in the literature; for example, see Pearce *et al.*, *Blueprint for a Green Economy.*

15 See S. Schmidheiny (ed.), *Changing Course*, MIT Press, Cambridge, MA, 1992.

16 This index was prepared for and is discussed in more detail in T. Jackson and N. Marks, *Measuring Sustainable Economic Welfare – a pilot index: 1950-1990*, Stockholm Environment Institute, Stockholm, 1994.

17 This is by no means the only way of making the judgement. Some have argued that it is more appropriate to consider several different indices

relating to different elements of welfare such as air quality, water quality, infant mortality, longevity, mineral reserves and so on. See, for example, Victor Anderson's *Alternative Economic Indicators*, Routledge, London, 1991. In 1991 the OECD published a 'preliminary set' of environmental indicators for the OECD countries.

18 The US index was developed by Herman Daly and John Cobb in their book *For the Common Good – redirecting the economy towards community, the environment and a sustainable future*, Green Print, London, 1990, and by Clifford Cobb and John Cobb in *The Green National Product*, The Human Economy Center, University Press of America, Lanham, MD, 1994. A similar index for the UK was developed by Jackson and Marks in *Measuring Sustainable Economic Welfare,* and an index for Germany developed by Diefenbacher is also published in Cobb and Cobb, *The Green National Product*.

19 L. Summers, 'Research Challenges for Development Economists', *Finance and Development*, vol. 28, pp. 2–5, Washington, DC, September 1991.

20 See the introduction to the World Resources Institute's *World Resources 1993–4*, Oxford University Press, Oxford, 1994.

21 These figures come from the World Resources Institute's *World Resources 1992–3*, Oxford University Press, Oxford, 1992, p. 19.

22 Some would also add war and famine to its list of crimes. Douthwaite, for example, makes a strong case that the capitalist development of Western Europe led directly to the First World War – with the loss of almost 9 million lives. Schumacher also makes the case strongly in *Small Is Beautiful*. The evidence that Britain's rapid industrialisation was based in part on military might is also compelling. And resource domination has been closely implicated in more recent military conflicts, such as the Gulf War. 'War is a judgement', wrote Dorothy Sayers, 'that overtakes societies when they have been living upon ideas that conflict too violently with the laws governing the universe.'

10 BEYOND MATERIAL CONCERNS

1 The results of the survey are quoted in Douthwaite, *Growth Illusion*.

2 Cited in M. Lutz and K. Lux, *Humanistic Economics – the new challenge*, Bootstrap Press, New York, 1988.

3 Indeed, it is not difficult to find this kind of argument even today. Wilfred Beckerman in his book *Pricing for Pollution*, Institute of Economic Affairs, London, 1990, bases his entire argument on the 'timelessness' of the principle of economic self-interest.

4 See, for instance, L. Doyal and I. Gough, *A Theory of Human Need*, Macmillan, London, 1991, and references in it; M. Max Neef, A. Elizalde and M. Hopenhayn, *Human Scale Development – conception, application and*

further reflections, Apex Press, New York, 1991, and P. Ekins, M. Hillman and R. Hutchison, *Wealth beyond Measure – an atlas of new economics*, Gaia Books, London, 1992, also discuss these issues.

5 This is really the direct translation of Bentham's unidimensional pleasure calculus into the realm of the market.

6 The list of authors who have also dismissed this equation is both long and impressive. It includes Mill himself, Sismondi, John Ruskin, E.F. Schumacher, Abraham Maslow, Carl Rogers and many others. Lutz and Lux's book, *Humanistic Economics*, provides a fascinating history both of mainstream economic development and of the humanist response to it.

7 Incredibly, some economists have attempted to argue against the distinction between wants and needs, even where such basic needs as hunger and thirst are concerned. 'Do we need water?' asks Paul Heyne in *The Economic Way of Thinking*, Chicago, 1967. 'No. The best way to turn a drought into a calamity is to pretend that water is a necessity.'

8 See, for instance, *Motivation and Personality*, Harper & Row, New York, 1954.

9 These needs were also known as self-actualisation needs.

10 See *Towards a Psychology of Being*, Van Nostrand Reinhold, New York, 1968.

11 See Max Neef *et al.*, *Human Scale Development.*

12 This need was also translated as 'leisure' in the original project. The concept of leisure as a need is probably more acceptable to a Western view than the concept of idleness!

13 Actually, the project described economic goods as the material 'manifestation' of satisfiers, and defined five different categories of satisfiers: synergic satisfiers, singular satisfiers, pseudo-satisfiers, inhibiting satisfiers and violators. My use of language here is slightly different. But the categorisation is roughly the same.

14 See Table 2 in Max Neef *et al.*, *Human Scale Development.*

15 See in particular an excellent discussion of alternative agricultural techniques in P. Goering, H. Norberg-Hodge and J. Page, *From the Ground Up*, International Society for Ecology and Culture/Zed Books, London, 1993.

16 This can be characterised in the Max Neef framework as an interacting participation need.

17 Community itself is not expressed as a need in the Max Neef framework. But it can be seen as a satisfier of the needs for participation, affection, identity and perhaps also protection.

18 Again, there is no 'need' for spirituality in the Max Neef framework. But there are certainly resonances here with the suggested tenth need of transcendence, and with Maslow's self-actualisation needs.

19 L. Herber, *Our Synthetic Environment*, Jonathan Cape, London, 1963, cited in Schumacher *Small Is Beautiful*, p. 94.

SELECT BIBLIOGRAPHY

Campbell, M. and Glenn, W., 1982, *Profit from Pollution Prevention*, Pollution Probe, Toronto.

Coveney, P. and Highfield, R., 1991, *The Arrow of Time*, Flamingo, London.

Daly, H. and Cobb, J., 1989, *For the Common Good – redirecting the economy towards community, the environment and a sustainable future*, Green Print, London.

Doyal, L. and Gough, I., 1991, *A Theory of Human Need*, Macmillan, London.

Debeir, J.-C., Deleage, J.-P. and Hemery, D., 1991, *In the Servitude of Power: energy and civilisation through the ages*, Zed Books, London.

Douthwaite, R., 1992, *The Growth Illusion*, Green Books, Bideford, Devon.

Ekins, P., Hillman, M. and Hutchison, R., 1992, *Wealth beyond Measure – an atlas of new economics*, Gaia Books, London.

Georgescu-Roegen, N., 1971, *The Entropy Law and the Economic Process*, Harvard University Press, Cambridge, MA.

Goldsmith, E. and Hildyard, N., 1992, *The Earth Report 3: an A–Z guide to environmental issues*, Mitchell Beazley, London.

Hirschhorn, J. and Oldenburg, K., 1991, *Prosperity without Pollution: the prevention strategy for industry and consumers*, Van Nostrand Reinhold, New York.

Hobsbawm, E., 1968, 1990, *Industry and Empire*, Penguin Books, London.

Jacobs, M., 1991, *The Green Economy*, Pluto Press, London.

Jackson, T. (ed.), 1993, *Clean Production Strategies: developing preventive environmental management in the industrial economy*, Lewis Publishers, Boca Raton, FL.

Lutz, M. and Lux, K., 1988, *Humanistic Economics – the new challenge*, Bootstrap Press, New York.

Max Neef, M., Elizalde, A. and Hopenhayn, M., 1991, *Human Scale Development – conception, application and further reflections*, Apex Press, New York.

Odum, E.P., 1983, *Basic Ecology*, Saunders College Publishing, Philadelphia, PA.

Schmidheiny, S. (ed.), 1992, *Changing Course*, MIT Press, Cambridge, MA.

Schumacher, E.F., 1974, *Small Is Beautiful*, Abacus, London.

US Bureau of Mines, 1991, *The New Materials Society: material shifts in the new society*, vol. 3, US Department of the Interior, Washington, DC.

US OTA, 1986, *Serious Reduction of Hazardous Wastes: for pollution prevention and industrial efficiency*, Office of Technology Assessment, US Congress, Washington, DC.

WCED, 1987, *Our Common Future*, report of the Brundtland Commission on Environment and Development, Oxford University Press, London.

INDEX